D1609245

A LOGBUILDER'S HANDBOOK

A Logbuilder's Handbook

Drew Langsner

1982

Rodale Press, Emmaus, Pennsylvania

Printed in the United States of America on recycled paper containing a high percentage of de-inked fiber.

Design by Linda Jacopetti
Illustrations by Gene Mater
Page 155 art based on an illustration courtesy of *Fine Homebuilding* magazine.

Library of Congress Cataloging in Publication Data

Langsner, Drew.
A logbuilder's handbook.

Includes index.
1. Log cabins—Design and construction.
I. Title.
TH4840.L36   694′.2   82-3857
AACR2

ISBN 0-87857-416-6 hardcover
ISBN 0-87857-419-0 paperback

2 4 6 8 10 9 7 5 3 1 hardcover
2 4 6 8 10 9 7 5 3 1 paperback

CONTENTS

PREFACE

For many years, my knowledge of log construction was based entirely on the popular image of the American log cabin—a rough dwelling hastily constructed in the wilderness by frontier families with little more than an axe and a team of oxen. A year-long tour of the European conifer belt, however, which includes Scandinavia, the Alps and parts of the Balkans, thoroughly changed my impressions and greatly increased my appreciation of the log house. I discovered that crafting fine log buildings was a time-honored tradition in Europe, ranking equally with such recognized crafts as furniture making or fine textile weaving. In the countries I visited, master logbuilders created beautiful structures that have been in continual use for centuries. A log house in these areas has long been considered an investment in the future.

Perhaps the most important quality that I began to appreciate was that many of the mountain folk live in close harmony with nature. One young Swiss farmer, for instance, said I could camp in his woods, but quickly added that I must be careful not to injure the pine seedlings that were growing there. The forest I was in was planted by his father when my host was a youngster.

I was also impressed with the care and planning that went into each building. I learned that local methods had evolved over many generations. That the builders were closely in tune with the resources and needs of each locality was clearly evident. The level of craftsmanship was awe-inspiring.

In Switzerland, I passed through village after village in which every inhabitant lived in an immense *Bauernhaus* (a farmhouse that combines home and barn in one building), each one beyond my wildest fantasy of what was possible with log construction. These multistoried structures, often masterpieces of architecture and woodworking, are enormous. A typical *Bauernhaus* includes living quarters for the farm family, a separate apartment for older parents, rooms for summer help, stalls for five to ten dairy cows, a horse stall, workshop, root cellar and a gigantic loft for storing farm implements and loose hay.

By contrast, the traditional farmstead of Scandinavia consists of a group of log buildings, smaller than the Swiss structures, and each with an individual purpose. The most usual buildings are the *stue* (home), cow barn, stable, *loft* (granary), sauna and storage shed. Often all of these are arranged as a compound with a connecting fence between the structures. Scandinavian log buildings combine simplicity and elegance. The northern axeman loved the warmth and powerful strength of immense logs, and though the typical Scandinavian log building is small, the logs with which it is built are sometimes over 2 feet in diameter.

After seeing what could be done, I got the urge to try my own hand at logbuilding, and in 1974 my wife Louise and I moved to western North Carolina, to a mountain farm just below the Appalachian Trail. For the first few years we lived in an old "double-box" cabin (a frameless structure consisting of alternating inch-thick boards nailed back to back) that came with the farm, but in 1978 we started building our first log house. By then, I had logbuilding fever!

In planning our house, Louise and I wanted not only to combine various features of both European and American log buildings, but also to preserve something of the feeling of each style of architecture as well. It was also important that what we built be in harmony with the traditional buildings of our region, which is one of the few places in North America where log buildings are still commonly seen. Some of our neighbors grew up in log houses, and many of the older men here are excellent axemen.

We decided to build a two-story structure with a large gable roof that would protect not only a porch and balcony but also the log walls and the corner notches. We chose to follow the Scandinavian tradition of building a group of small structures, since we didn't want to take on all at once a project that was larger than our capabilities. Rather arbitrarily, we chose 20 feet as the maximum interior dimension for our house. Figuring a 6-inch wall thickness and 3 inches of notch extension, the longest logs would have to be at least 21 feet, 6 inches in length. (Heavy enough for us!) About 100 feet away we planned a smaller, multistory storage building, or "annex." The barn, woodworking shop and storage sheds would come later.

That year we also began our crafts school, Country Workshops. An experienced homesteading neighbor, Peter Gott, agreed to teach a logbuilding class, which helped us get started on the house. Although broadaxe work and log notching is slow, especially for beginners (at first it took me most of a day to hew and notch a log), during the weeklong class the floor joists and the first course of logs were cut and fitted, and the participants, myself included, learned the basic skills.

Building the stone foundation and log shell required a full year's time squeezed among all the other duties of homestead living. The next year work was temporarily interrupted when Daniel O'Hagan, another friend who has been building log structures for years, agreed to teach a logbuilding workshop. We had to take the time to build another foundation (for the annex, as it turned out), do more logging, and prepare for the class itself. Originally we had hoped to get the house roofed by the second winter, but splitting shingles was more time-consuming than expected (although very enjoyable work). When snows came we were forced to halt

Photo p.1 *Peter Gott hewing.*

Photo p.2 *The Gotts' house. There is a living room downstairs and a bedroom upstairs. The shed-roof addition on the right is the kitchen. Two children's rooms are in the addition on the left.*

that year for good, since we were unable to work safely on the roof. The last shingles were laid during the spring of the third year. During the months that followed, we chinked the logs (another big job), put in flooring, insulation, one door, windows, and the stainless steel flue for the stove.

The old people of our mountains say that "a log house is the warmest house there ever was." Compared with old-fashioned double-box

Photo p.3 ***The Gotts' storehouse. Two freezers and home-canned provisions fill the ground floor. Peter's woodworking shop is upstairs.***

Photo p.4 ***Combination barn and studio. Stalls for a family cow and calf are at ground level. Polly Gott's pottery and painting studio occupies the log pen. The loft is for hay and summer guests.***

construction, this is true; however, a carefully designed modern house, built with today's techniques, is more energy-efficient than either log or box. How can I justify building a log house in the 1980s? For one thing, our house is small, about half the size of an average contemporary home. For another, our farm is energy self-sufficient. We heat and cook entirely with waste wood from our own place. We've even incorporated several passive solar features to reduce the amount of wood burned each year. Really, though, my justification for building with logs is simply that I wanted to know the feeling of wielding a broadaxe and an adz, of hewing and notching logs by hand, and of building a house that belongs in tradition and spirit not just to our farm or our region, but to the entire heritage of craftsmanship that finds its expression as much in the American owner-built home of today as in the pioneer homestead of the past, and in their European and Scandinavian antecedents of long ago.

When I started writing, I quickly realized that this handbook must be limited to those aspects of logbuilding with which I am personally familiar. I decided to include details on the processes which I have used, with some related material that should be helpful to other logbuilders. I have tried to distinguish between personal knowledge and ideas garnered through my research. I am well aware that logbuilders are great individualists and that most of us have something unique to contribute to the craft.

I am particularly grateful to my logbuilding teachers, Peter Gott and Daniel O'Hagan. Peter has developed techniques for chinked and hewn logwork that have raised the standards of American logbuilding to a new plateau of craftsmanship. The technique of hewing presented in this book was developed by him during nearly 20 years of logbuilding. Once, during a workshop, Peter said that "the goal is to do a rustic but perfect job." Peter continues to teach and do custom logbuilding, in addition to calling square dances, playing in his family's old-time country music band, and tending to the myriad duties of homestead living here in Marshall, North Carolina.

Daniel O'Hagan also has been a great inspiration. He is someone who demonstrates in everything he does the possibility of living by one's personal beliefs. Daniel's approach to building is direct and artistic—his methods are more poetic than technical. Daniel once said, "I like to do things the hard ways, because it is better."

Our home and storage building would not have been built without the help of many devoted friends. John Chiarito shared his keen knowledge of building during his annual visits from California. Sian Newman-Smith proved to her own satisfaction that a woman can hew and notch logs—she did more than a few of ours. Other friends and relatives have helped with log raising, chinking, nailing shingles, hanging rain gutters and countless building details.

I would also like to thank my friends at Rodale Press. Managing Editor Bill Hylton encouraged me to start this book. John Blackford did a great deal of work smoothing out my manuscript. John Warde made sure that the chapters on hewing, notching and roofing were comprehensible, in addition to overseeing the editing and general production.

And of course Louise has been here right along. Her work began when I started dreaming of building.

CHAPTER I

HISTORY

Many architectural histories focus on the residences of upper classes, overlooking the folk residences of the same period. Yet it is the buildings of the common people—their dwellings and storehouses—that have a real story to tell. Folk architecture exhibits the subtle relationship between shelter and the owners' conception of their place in the universe. The structures invariably reflect the environment that has nurtured them.

Many different construction methods have arisen more or less simultaneously in different geographical regions. Mixes then occur as a result of migration, travel and trade. It's not uncommon to find several distinct construction techniques in one building or a group of buildings constructed at the same time—physical evidence of mingling styles. The wooden folk houses of Europe were usually of either post-and-beam timber-frame or log construction. Post-and-beam structures consist of vertical and horizontal timbers, usually supplemented with diagonal bracing. Typically, the timbers are connected with various types of pinned mortise-and-tenon joints to form the erect frame. The spaces between the timbers are then filled in with stone, brick, wattle and daub, or other materials. Filled-in post-and-beam construction such as this is usually referred to as half-timber architecture.

The structures employing post-and-beam construction were found in the deciduous forests of western Europe. Log construction, on the other hand, which consists of stacked horizontal logs locked together by corner notches, flourished in the evergreen forests of central and eastern Europe; the long, straight tree trunks available there facilitated the technique. Logbuilding may in fact have originated in Russia and spread from there to Scandinavia, also noted for forests of tall conifers. Logbuilding techniques are believed to be somewhat more recent than post-and-beam methods, but the origins of both types of construction are lost in the dark forests of a prehistoric period.

SCANDINAVIAN LOGBUILDING

The Norwegians developed their own distinctive logbuilding tradition, grafting techniques from boat building and from a type of timber framing called *stave* construction to the Russian style, which consisted exclusively of round-log construction. Characteristic Norwegian buildings have wall logs hewn to an oval cross section, with projecting joints at the corner and curved cantilevers supporting balcony and roof overhangs. The straight, uniform trees of the northern forests were particularly well suited to the logbuilding style of the Russians and Scandinavians, who fitted their logs together so tightly that no space was visible between them. The superior fit was achieved by hollowing a lengthwise channel along the bottom of each log, which exactly matched the contour of the log below. Moss was tamped into the channel to seal off any air leaks. This method, called chinkless or scribed-log construction, has been revived and modernized by the contemporary Canadian logbuilder, B. Allan Mackie, whose books on the subject are listed in Appendix V.

The basic Norwegian log building was typically expanded by the addition of porches and balconies joined to the core log structure by means of shared sills and plates, as well as by logs that extended past the house wall to form the sides of the addition. At the Norsk Folk Museum in Oslo, Norway, reconstructed log buildings duplicate the layout of old farms throughout the country. The arrangement of the buildings varied from irregular clusters in western Norway, to rows in the central region, to square enclosures in the east. The log traditions of Sweden and Finland also incorporated separate unit construction, but with layout organized to permit walls to join the units, forming a protective enclosure.

The typical single-floor Norwegian log *stue* contained three rooms with overall dimensions occasionally as large as 26 by 33 feet. The home was dominated by a central, open hearth that filled the interior with

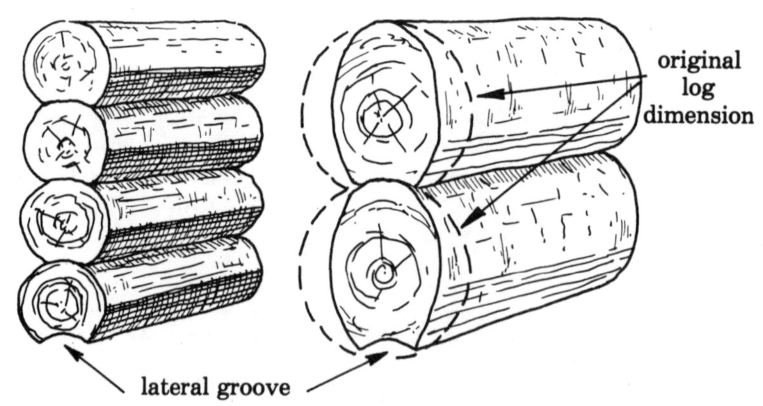

Illus. 1.1. ***End view of Scandinavian round-log construction with lateral grooves. In Norway, very large logs were sometimes hewn to an oval cross section.***

Photo 1.1 ***A Norwegian forest hut built with scribed round logs and a sod roof. Reconstructed at Lilliehammer, Open-Air Museum, Norway.***

smoke. Chimneys did not become common until the 17th century, when Finnish immigrants introduced enclosed-chimney ovens with a hearth for cooking. The main room was used for food preparation, eating and sleeping, and it was furnished sparsely, usually with only built-in wall benches, a table and a box bed. There was also a back chamber for storage and an entry room, which joined a half-enclosed front porch facing the farmyard. During the Middle Ages these log houses seldom had windows.

Unlike the ground-hugging, straightforward *stue,* a Scandinavian *loft* rose two stories, its jettied second floor often presenting an ornately carved facade. Its airborne image was intensified by the relatively short length, roughly 16 to 20 feet. Old-fashioned lofts functioned as the farm treasury: On the ground floor grain, meat, bread and other provisions sat in large kegs and vessels, while the upper floor held special clothes and valuables. The loft also doubled as a summer gathering place.

Sod over layers of birch bark was a common roof style in old Norway, as was slate. In summertime, sod roofing looks beautiful, with its tall green grasses, ferns and wild flowers in bloom. Sod roofing is fire

Photo 1.2 ***A classic three-room dwelling house combining notched-log and post-and-beam construction. Located in the Numedal Valley, Norway.***

Photo 1.3 ***Two storehouses; Numedal, Norway.***

resistant and provides fairly good insulation, though slate is more durable, good for a century or more. In Finland, thatch was often used as a roofing material. Thatch has excellent insulation value and can last up to 75 years, though dry thatch is extremely combustible. Roof framing for a thatched roof must be very strong because thatch is heavy when it becomes water-

Photo 1.4 *A round-log barn, probably constructed in the 15th century; Diemtigal, Switzerland.*

logged. Rows of hollowed, half-round poles with overlapped channels fitted together like long cylindrical tiles were also used as roofs in Finland. After machine-made nails became available, shingles split from pine became the standard roofing material. These last for 20 to 40 years. Ceramic tiles also came into use, and today many old log buildings in Norway have been reroofed with tiles of fired clay.

To me, the most impressive aspect of Scandinavian logbuilding is the spiritual and creative sensibility that dominates the architecture. The Norwegian logbuilders had a great sense for emphasizing the three-dimensional shape and mass of logs and the organic nature of wood. The bond between men and trees from ages past is clearly transmitted through these houses—houses created by a people whose image of the cosmos was, according to Norwegian architect Christian Norberg-Schulz, that of an immense tree.

Photo 1.5 ***A small storehouse combining notched-log and post-and-beam construction; Diemtigal, Switzerland.***

SWISS LOG HOUSES

The Swiss Alps contain a wonderland of wooden-house styles. These central European mountains have played an historic role as a cultural crossroads for centuries. At the same time, Switzerland contains hundreds of isolated valley communities, each relatively distinct and independent. Conservative in nature, these alpine communities have preserved their individual identities while slowly assimilating various elements of building style from neighboring regions, resulting in a wide variety within a distinct tradition.

Log construction in Switzerland seems to have begun with the use of round logs notched together for animal shelters and hay lofts at a time when half-timber architecture was already highly developed for the construction of dwelling houses. A few round-log barns built 500 to 600 years ago are still used. The adoption of log construction in the Alps represented an improvement over the older tradition of post-and-beam framing because the insulation value of solid log walls is far greater than the combination of timber and masonry fill.

Photo 1.6 *Keyed and dovetail notches on a Swiss farmhouse. The logs were squared on four sides at a sawmill, then hand-planed smooth.*

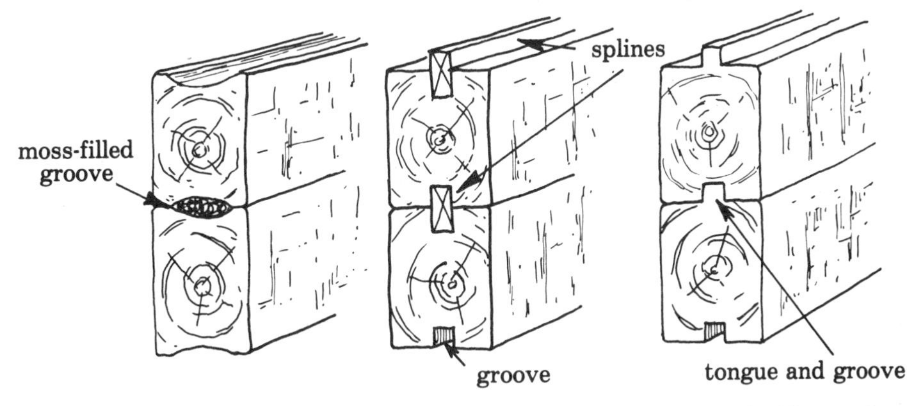

Illus. 1.2. *Early Swiss hewn-log construction used a shallow groove filled with moss for a weatherproof fit. Milled logs used a fitted spline. Modern versions are tongue and groove.*

Photo 1.7 ***A herder's summer quarters built into a slope high in the Swiss Alps.***

By the early 18th century, it became the custom to saw logs to a square or rectangular cross section. In some cases logs were sawed in half lengthwise to get two timbers out of one log. The log sides were then smoothed with large, two-man hand planes. Early sawed logs had a cavity running lengthwise, on the top and bottom of the timbers, which was filled with moss in the Scandinavian manner. By the 19th century, the cavities became sawed grooves with fitted wooden splines instead of moss. As log-shaping machinery developed, tongue-and-groove methods were developed.

In the Alps, one finds log construction carried to technical perfection. Logs were squared on the sides to simplify layouts and geometry, and to allow greater precision. Builders created long walls by incorporating post-and-beam techniques, fitting the ends of shorter logs into slotted vertical posts. The same techniques were used to ensure the rigidity of the wall when openings for doors or windows were made. Logs at the opening were mortised into slotted posts. This method allows for expansion and contraction owing to moisture variations, since the horizontal logs are not anchored to the vertical door or window frame. The flooring system was also planned with consideration for seasonal changes in dimension caused by varying moisture content. Central, tapered floorboards, for instance,

Photo 1.8 ***Chalet constructed in the 18th century. The original shingled roof has been covered with ceramic tiles; Diemtigal, Switzerland.***

often protrude through the wall of a house; the boards are forced outward as the other tongue-and-groove floorboards expand. During dry seasons when the wood shrinks and the floor needs tightening up, the wedge-boards are pounded back into place.

I would single out the chalet style of Swiss house as one that represents the highest achievement of Switzerland's log tradition and is symbolic of the country's housebuilding art. The word chalet derives from the French *châtelet,* meaning small castle. The chalet was, however, developed as the house of free peasant farmers, whose tradition of independence goes back as early as the 13th century. The chalet (a type of *Bauernhaus*) housed both people and farm animals under one roof. In fact, space for all aspects of farm life—residence for both owners and helpers, food and equipment storage, animal shelter and workshops—was found within this one building.

The most remarkable feature of a chalet is the large size of the gable wall, which may be as much as 60 feet across, compared to 45 feet along the eaves. The resulting roof line is relatively shallow, but enormous; the eaves may extend as much as 10 to 15 feet beyond the gable facade, 6 to 10 feet beyond the lateral walls, and 3 to 6 feet past the rear wall. The area covered by the eaves is almost as great as that within the walls. These overhangs are supported by cantilevers of extended corner or interior wall logs. The openness of the gable is emphasized by the rather low height of each floor, 7 feet on the average. Usually the chalet is built into a slope. The

Photo 1.9 ***A Swiss valley*** **Bauernhaus.** ***The massive gable roof is hipped at each end, with a dormer over the eaves sides.***

"face" of the house, the gable facade, is traditionally oriented to overlook the valley below. This front wall is sometimes highly ornamented, its inviting character enhanced by almost continuous rows of windows above boxes of red geraniums that bloom throughout the summer.

The dwelling area is divided into two distinct parts by a central axis defined either by the roof ridge or by a line running midway between the front and back, perpendicular to the ridge. Living rooms and bedrooms have a southern or southeastern exposure, usually forming a compact cluster on two floors, with the kitchen in the rear to provide easy access to the outside service facilities, storehouse and cellar. The kitchen ceiling on older buildings formed a pyramid that extended through the upper story to a smoke hood in the roof. Partition walls lock into the outer walls at right angles, with logs extending on through to function as cantilevers to support the overhanging eaves outside.

The fine traditional heating system of Swiss houses is the warming oven, a heavy masonry stove with an interior baffle system. The stove was usually placed in the living room or bedroom, but with the firebox opening in the adjacent kitchen or corridor. In this way, the heated room was kept free of drafts, smoke and debris. A warming oven requires a lot of floor space, but is a wonderful thing to sit near during cold winter weather. Because of its thermal mass, it needs firing only once or twice every 24 hours. An early version of the stove was simply an immense masonry firebox protruding into the living or sleeping room but fed from the adjoining kitchen. Smoke circulated over a shelf and exited through a hole over the oven door, back into the kitchen (a very black place). Later, warming ovens acquired more elaborate baffle systems and the smoke was channeled to a chimney housed in an interior wall.

THE FRONTIER TRADITION: AMERICAN LOG CABINS

Although rooted in European wooden-house traditions, the log house as it appeared in the wilderness of North America was a unique hybrid, imitating none of its forebears. The colonists of Fort Christina in the colony of New Sweden on the Delaware River built the first log houses in America around 1640. Earlier colonists—the British in Jamestown, Plymouth and St. Marys, and the Dutch in New Amsterdam—built some log fortifications, but they were apparently not tempted to construct log houses. The early British settlers clung to the timber-frame customs of their homeland, rather than experiment with log buildings. New Sweden was settled by Scandinavians (though often identified as "Swedes," approximately half of the original settlers were Finns), who brought with them a knowledge of forestry and superb skills as axemen. An awestruck Englishman observed that his Scandinavian neighbors could "cut down a tree and cut him off when down sooner than two [English] men can saw him."

The log houses of New Sweden were primitive. Even though logbuilding in Europe was a highly developed craft practiced by specialists, many settlers came from the lower classes and were accustomed to rough accommodations. The peasant farmers arriving in America had few tools, and required shelters without delay.

The first log houses in New Sweden were small structures with low doorways; it was necessary to stoop to enter. Small holes covered with board shutters served as windows. The floor was dug in about 2 feet below ground level; hewn, round and half-round logs were all used in different dwellings. Chimneys were of stone, or of sticks and clay.

As a political entity, New Sweden was short-lived. But the building methods introduced there took root and spread. By the 1670s, Swedes and Finns were crossing the Delaware to settle permanently in New Jersey and Pennsylvania. Thus, a relatively small number of Scandinavian settlers were instrumental in establishing the log house in America.

Further development and dispersal of this frontier dwelling was largely due to two other migratory waves from Europe. In 1681, William Penn published "Some Account of the Province of Pennsylvania," which was circulated in Europe and aroused the interest of several Protestant groups. Quakers, Mennonites, Amish, Moravians and other sects were eager to try life in the New World. In America, they became known as the

"Pennsylvania Dutch" (from *Deutsch,* meaning German). Many of these immigrants were skilled farmers and artisans, familiar with the log construction of their native lands. They readily adopted the quick, chinked-log style pioneered by their Scandinavian predecessors. Although many houses of undressed logs were raised, most of the Germanic settlers hoped to replace these initial shelters with square-hewn logs, if not brick and stone. As the population of Pennsylvania grew, so did the number of log houses. Soon the Pennsylvania Dutch were moving on to Maryland and Virginia, and later to Ohio and North Carolina, spreading the log house as the typical frontier dwelling.

The third major contribution to the American log house came from the Scotch-Irish. Although they had no background in log construction, the newcomers rapidly adjusted to the demands of the American environment and imitated the houses devised by the earlier immigrants. Many of the early Scotch-Irish cabins were crude, relying on a hole in the roof rather than a chimney, with dirt floors and few, if any, windows. Like the Scandinavians, many of the Scotch-Irish came from poor backgrounds, but as they gained experience, their craftsmanship improved. They incorporated carefully notched hewn logs, and stone chimneys into their permanent dwellings. The exterior chimney on the gable wall, a standard feature of the American log house, appears to be largely due to Scotch-Irish influence.

Frontier cabins tended to be very small. Many of the earliest were huts built by trappers, or temporary shelters for use while permanent housing was being built. Wall height was sometimes as low as 4 or 5 feet; roofing often consisted of bowed, lashed saplings and limbs covered with bark. These "turtle backs" were soon abandoned for better quarters, to become hog houses or chicken coops before they crumbled back into the earth.

Perhaps the most famous American folk house is the one-room log cabin, built with round or hewn logs and typically no more than 16 by 20 feet. Carefully crafted examples have survived for well over 100 years. The shingled roof was usually a conventional gable with a steep pitch to shed water and to make room for a sleeping loft above the ground floor. Shed additions under the eaves were either enclosed for a kitchen or left open—creating the American front porch.

Other log houses were expanded by add-on units. Separate rectangular additions (called pens) to enlarge the house offered several advantages: The house could be built in stages, logs remained within reasonable size for handling, and rigid walls were ensured. A second pen was sometimes built directly against the gable wall of the first or erected several feet away and joined by a common roof. This latter design is named for the covered space between the two pens, called the "dog trot."

Logbuilding was introduced on the northwest coast by Russian fur trappers settling in Alaska. By 1812, the Russians established a colony in northern California, now known as Fort Ross, and built fortifications and residences there with round and hewn logs of redwood.

In Europe, the development and evolution of logbuilding can often be traced in both local communities and generalized regions, with distinctive, homogeneous building types closely identified with various ethnic groups. But early America was a frontier, characterized by mobility and unexpected meetings of various cultures. The American log house was a hybrid developed to suit the needs of settlers who required shelter quickly. Sawmills were usually close behind the frontier, and as a result, timber-frame and lumber construction often took over before local logbuilding methods had a chance to mature. Mechanized nail production also revolutionized building methods.

Nevertheless, there are several typical characteristics in American log construction. The most prominent is that open spaces left between logs were usually chinked. Tightly fitted logs, of course, were packed with

Photo 1.10 ***Classic American logwork combines hand hewing with clay chinking between log courses. Reconstruction at the Vance Birthplace, near Weaverville, North Carolina.***

Photo 1.11 ***A corn crib and storage loft at the Vance Birthplace.***

moss, but wider spaces were filled with sticks plastered with various mixtures of clay, lime and chopped straw. In some areas, such as Maryland and parts of Pennsylvania, log buildings were often designed to be sheathed with clapboards in the future, at some point when time and materials became available. In cases such as these, the bands of chinking could be as much as half the log diameter. Other typical American characteristics are the location of an open fireplace at the midpoint of one gable wall and the placement of the main entrance at the center of an eaves wall.

Although some of the surviving examples of early American log houses show a high level of craftsmanship, logbuilding in this country has remained primitive in comparison with the European achievement, where farms were handed down for generations, and the family house was often centuries old. From the pioneer's point of view, America was a wilderness, but one that encouraged innovation. There was nothing to hand down, few cultural dictates and plenty of trees. The log house in America represents the heritage of the owner/builder—a little rough, but serviceable, and built with pride and enthusiasm.

CHAPTER II

LOG-HOUSE ECONOMICS

The true cost of any house extends far beyond simple cash outlay. To begin with, you should include the total energy input that goes into production and transportation of commercial building materials, plus the many factors supporting the labor force necessary to manufacture and assemble the building. There are also design decisions that effect thermal efficiency, which in turn determines the amount of fuel used for heating. Even the longevity of a structure, its potential resale value and maintenance costs all influence its ultimate worth.

The relationship of a dwelling to the environment reflects the homeowner's sense of responsibility as a participant in the social and ecological community at large. In contrast, "bottom line" and "supply side" economics tend to reflect the situation of the moment, with limited regard for the future or the general environment. An owner-built log home, built with local materials, requires no large expenditures of fuel to transport the materials across the country and no charges for dressing and kiln-drying the wood, so the cost in both cash and resources is less than for most frame buildings. You eliminate distant middlemen while helping the local economy. If you manage your own woodlot carefully, the logs for your home are a sustainable resource, available without damage to the surroundings.

Building with trees from your own property is beneficial because forests are now being depleted faster than they can be replaced. Where does the wood go? In the United States in 1977, 57 percent went into pulp and paper products, 24 percent for lumber, 10 percent for fuel and 9 percent for plywood, particle board and miscellaneous uses. Logging operations commonly harvest only 50 percent of the potential wood in an average tree. Many sawmills waste another 50 percent; it is common that only 25 percent of the wood of a complete tree actually becomes lumber. Log buildings contain more cubic feet of wood than conventionally framed houses, but there is much less waste than with lumber, and a properly built and maintained log house should outlast a frame structure by a substantial margin.

Even if you don't have enough timber for a log home, you may still be able to bypass the large-scale wood-processing industry by purchasing timber from a local source. It's true that these large-scale enterprises are now able to use much of the waste from logging and sawmill operations, producing paper products, building materials and wood-based synthetics, but the huge capital investment needed to set up these plants requires their continuous operation. The modern lumber industry consumes an enormous volume of material, which can only be supplied by extensive—even overextensive—tree harvesting. In addition, these "super-efficient" operations are often major contributors to soil, water and air pollution. High-technology logging and wood processing is also energy-intensive.

In contrast, logbuilding is a labor-intensive activity. Several dozen

Photo 2.1 ***Log-house economics are based on use of locally available materials.***

trees and a truckload of lumber can provide for most of a season's work for one or two skilled craftsmen. Waste and pollution from logbuilding are easily kept to a minimum. The small quantity of logging slash can be stacked into wildlife shelters that gradually decompose to rebuild forest soils. Tree tops can often be used for joists or rafters. Laps and other waste can be saved for firewood. (We cooked and heated with hewing chips for almost two years.) Sometimes tree species or woodlots with marginal commercial value can provide excellent material for building a log house. At their best, log buildings can represent the legacy of the forests.

Economic involvement can also be kept at a local level.The flooring and other lumber for our own house was sawed and planed at local mills. A neighboring farmer carefully skidded the logs out of our steep woodlot with his tractor. Total investment for logging and building tools was well under $1,000, and that money went to local merchants rather than huge national chain stores.

Do-it-yourself logbuilding offers another benefit, the great satisfaction of building your own home from scratch. Other examples of real handmade houses are those constructed of logs, stone and masonry, adobe, and post-and-beam timber-framing. In contrast, construction methods that depend on factory-made materials purchased from building supply centers never seem to achieve the same personal quality.

The actual cash outlay required for a log home is determined not only by the size of the building, but also by the cost per log and the extent to which you do the work yourself (or with the help of generous friends). If you own a suitable woodlot, the main construction material is nearly free. Still, the costs for windows, finish materials, appliances, a stove and whatever professional services you contract will add up rapidly. An owner-built log home may be a bargain compared to standard housing, but it can cost a lot. In fact, a fully contracted log home usually costs more than a comparable frame structure.

EXPENSES

We designed our log house as two separate structures, a basic three-room living area and a detached storage annex. The multistage plan allowed us to move into the house while we were still working on the annex, and spread out our expenses. By March 1981, the cash outlay for the 640-square-foot living area came to $4,831.00, or $7.55 per square foot. The annex and a bathroom addition will give us a total of 1,030 square feet. Costs per foot should remain about the same, or possibly decrease, since our water system is in, and there will be fewer windows in the annex. These expenses do not include land, logs or tools. The only costs for purchased labor were for a tractor operator to skid logs to the site and backhoe work for the water system. The house has no electricity. We use a compost privy and a very simple graywater system.

Table 2.1 **Costs for Langsner Three-Room House**
(At move-in time, fall 1980, still incomplete were porch, electric installations, and solar greenhouse.)

Item		Cost
Logging (tractor hire)		$ 180.00
Foundation (cement, sand, gravel, rebar, flashing)		392.00
Lumber		599.00
Sawing costs	$284.00	
Planing costs	99.00	
Purchased	216.00	
Roof		381.00
Shingle logs	285.00	
Gutters	96.00	
Hardware		514.00
Nails	151.00	
Miscellaneous	299.00	
Sander rental	64.00	
Water system (reservoir, backhoe hire, 600 feet of pipe, cold water plumbing, graywater system)		717.00
Windows (contracted) hinges		891.00
Chinking (metal lath, sand, cement, nails, color)		215.00
Insulation		268.00
Finishes wood preservatives, caulk		357.00
Flue (stainless steel, insulated)		299.00
		$4,813.00

Our house has been a low-budget project, but we still could have cut our costs considerably. For instance, we used all new materials. A fair amount could be saved with salvaged windows, hardware and finishing materials. We used wood preservatives made with purchased ingredients. Old motor oil could be substituted. For chinking we experimented with plasterer's metal lath and stucco filled in with strips of Styrofoam insulation. Traditional chinking, consisting of sticks, moss and clay, costs nothing.

However, expenses can skyrocket far beyond those in table 2.1, which shows our costs. Skilled labor does not come cheaply and neither do commercial materials. Electrical wiring and an official wastewater system would have added substantially to the costs of our home. The bill for any commercial building materials (for instance, window units and flooring) or built-in appliances can quickly add up. Some of the log palaces being built today are splendid examples of conspicuous consumption, but log houses are best suited to those on low budgets who want to take the time to do most of the work themselves. In these circumstances, log-house economics look quite attractive.

Careful site orientation, strategic window placement, and good insulation and weather stripping can easily be incorporated into a new log house. Good insulation—including roof or ceiling insulation—is always cost-effective; proper orientation and window placement costs nothing; and though double-glazing, movable window insulation and similar energy savers may add to your initial expenses, they will return the investment over time and improve your comfort immediately. Log houses also gracefully incorporate roof-mounted solar panels.

On one hand, the question of whether to build with logs involves personal finances: Do you have enough cash, credit and free time to make it happen? Is a log home a sound investment? On the other hand, the question involves less tangible issues: the long-term value of logs as a building material and the personal factor of building your own home with local materials. Any decision must weigh the larger questions against what's right for you and your family.

CHAPTER III

SITE SELECTION

Perhaps the most important single decision you'll make in designing a house is the selection of a building site. Site selection is particularly important in farming or homesteading, where you are directly dependent on the land for food, fuel, building materials and water. A well-sited house combines privacy with reasonable access. Most people like to combine a fine view (but not necessarily a panorama) with a feeling of coziness and protection. A spot that receives plentiful direct sunlight, especially during winter, will reward the occupants with warmth. It's also advantageous to have a natural shade cover by late afternoon during the heat of summer. A dependable source of pure, plentiful water is a necessity.

Logbuilders should also consider site characteristics that will affect the building process. How will logs be hauled to the area? Where can logs be stored? Is there a good, level work area? (If not, it may be necessary to do the hewing and notching at a different location, and raise the walls themselves after all the logs are notched.) What system will be used for raising the walls? The simplest method, pulling logs up pairs of skid poles (see chapter 9), requires a fair amount of work space on two sides of the log pen. If logs can be brought to the base of one wall, you could use a homemade crane with a block and tackle. Logbuilders working at difficult sites have even been known to set up "skyline" cables strung between trees, similar to rigs used by loggers working in steep mountains.

An absence of habitable dwellings on a particular piece of land can indicate problems with the site. Usually, there are good reasons why some areas are populated, while others remain empty. In mountainous regions, sunny south slopes often support successful farms while shadier north slopes nearby remain in woodland. Lack of water, extreme winds, unstable subsoils or difficult access are problems that can be difficult to overcome. In some cases, code restrictions may make building impossible or impractical. It's unlikely that you're the first person to look over a place, but your requirements and means may be different from someone else's. Don't rule out an unsettled spot; just find out if there is a real problem while you can still change your mind.

My first suggestion in choosing a site is to take your time. Louise and I lived in the old "double-box" cabin on our farm for four years before we decided where to build. We considered several locations, but none really

Photo 3.1 ***The site should include easy access with spacious work and storage areas.***

satisfied our requirements. One day we simply "discovered" the spot where we have now built. Not surprisingly, it was the site of an older dwelling that had been abandoned some years earlier.

Of course, not everyone can take years to select a building site. But you may be able to set eagerness aside and wait one full year before beginning construction. Use this period to study the site, so that you actually *know* the place. Visit the site often, during every season. Make a point to be there at different hours, particularly sunrise, late afternoon and sunset. Note the locations of sunrise and sunset, and the time they occur, at the winter and summer solstices. Nighttime visits are interesting. You may discover prominent city lights beyond the horizon. The world of sound is quite different at night than during daylight. Camping at the location is highly recommended. Take the opportunity also to carefully design your house, integrating personal requirements with the given factors of the location. In addition, this is a good time to further your research into building methods and to locate the various tools and materials that will be required.

Note aspects of the site that will affect the house design. Shadows can work for or against you, according to when and where they come. Check the direction of prevailing winds. Some locations, like ours, get winds from several directions. It's a good idea to investigate several potential sites at the same time. Check the hours of sunrise and sunset in each place. Carry a thermometer; you may be surprised to find a consistent difference of several degrees between locations only a few hundred yards apart.

A year-long study will also provide an opportunity to observe the water situation during wet and dry periods. Is there any risk of flooding? Do springs run throughout the year? Have water tested by the health department, and check the drainage situation. Does uphill runoff wash onto the site? You may need to control runoff, possibly even grading to change the contour of the site. Also, dig a hole to check for groundwater, which indicates a potentially unstable subsoil. A hole 2 or 3 feet deep should be adequate, unless you are planning a basement, in which case a deeper excavation is called for.

Another project during this study period is locating the route for any road that may be needed. Consultation with neighbors (who probably know the place very well) is helpful. Besides, developing a good relationship with your neighbors is one of the most worthwhile things you can do at the new site.

Most people enjoy beautiful vistas, rushing brooks and waterfalls. Uplifting scenery may be one of your reasons for living in the country or wilderness in the first place. But locations with grand views often make poor building sites. Construction on top of a hill or on the side of a mountain, for instance, can lead to added expenses; bringing electricity to such a site, for instance, is often wildly expensive, and part of the view may be marred by the utility poles and wires. Road maintenance is also a factor, and those costs will continue indefinitely.

A steep road may also get rutted or dangerously icy during cold, wet weather, and it can take years off the life of your car or truck. On the other hand, a hillside site may be ideal for incorporating a full basement under the house.

High building sites are often blessed with long hours of sunlight, though they are more exposed to weather. This is fine—even splendid—when the weather is good, but when the weather is bad, you really get it. Sound travels up hillsides nicely. On a quiet day, with just the right breeze, you may be able to hear the conversation of neighbors a quarter of a mile away. Be prepared also to hear trucks on a distant highway, chain saws, tractors and motorcycles. Friends of ours found themselves vicariously participating in the excitement of a cheering football crowd at a high school four or five miles away. (They started going to the games!)

A knoll or slight projection is often a good site. Sometimes one can be located that has a modest view, gravity springs and the potential for a basement at grade level without the disadvantages of sites in more difficult (albeit spectacular) terrain.

Unless you live in a very hot region, look for a site with open southern exposure. Year-round temperatures here will be considerably warmer than at nearby northern exposures, because of the additional sunlight. Eastern exposure is also advantageous, because the sun will strike the site early in the morning, providing quick warmth. Nothing is perfect, however, so choose with all factors in mind: A sunny east-facing site might be so much windier than a nearby sheltered cove that the advantage would be nil. Also, the midwinter and early spring sunshine that warms a well-exposed southern site can cause periodic thaws, resulting in a longer "mud season" than might be found in protected areas nearby that remain frozen.

Consider how you will develop your water supply. A site with potential for a simple gravity-powered system fed by an uphill spring is good, but don't choose a hillside site purely for that reason, since a long run of buried pipeline could cost more than a well somewhere else. In a broad, flat valley, you will most likely choose to drill a well and install an electric pump.

The major advantage of a valley location is the abundance of flat land, which provides considerable freedom to organize building clusters, farming fields, and woodlots. Another advantage is that there are often serviceable roads throughout the area. If you have to build your own road, constructing and maintaining a roadway through flat country or low hills is generally simpler than in mountainous locations. The negative side of having good roads, however, is that you may not remain isolated for long as you might have hoped.

The water situation in low country is often different from that in nearby highlands. Valley farms may have an abundance of irrigation water and a scarcity of drinking water at the same time. This is particularly true now, in places where agricultural chemicals leach into the groundwater. There is also the possibility of flooding. Don't take this lightly: The flood from only one bad storm could destroy your house completely.

For many owner/builders, a nearby garden site is also important. It's possible to have a successful garden almost anywhere, but you might have to do a vast amount of work improving one site compared with a better one nearby. Try to choose a location that's level or fairly flat. Most garden crops thrive on long hours of direct sunshine, but some afternoon shade is also beneficial, especially for cool-weather crops. Loose rocks can be removed, but they seem to come up year after year. Clearing large stumps may require a bulldozer. If you need to bring in a crawler for roadwork or site preparation, consider terracing the garden at the same time. A garden water system is necessary in areas subject to drought and an appreciated luxury anywhere else.

Finally, consider the effects of site modification on your house plan. In general, log houses look best in a natural setting. Major changes in the landscape may be inappropriate. On the other hand, subtle modifications can enhance a site, especially if the work is carefully planned and executed.

CHAPTER IV

DESIGN

In many ways, a log house is similar to a frame structure, but an understanding of logs as a building material should precede specific layout or other design considerations. Horizontal log construction differs from frame construction principally in that log construction consists of stacked timbers arranged in one or more rectangles, called pens. The pen corners are joined with locking notches cut into adjacent horizontal logs. There is no corner framing or triangulated bracing as in frame construction. Joists and beams, however, can help in the alignment and support of the log walls and roof framing. Intersecting walls can be used also, to brace long exterior walls and support joists and roof beams.

Round-log construction is effective where straight logs with minimal taper are available. Notches are usually cut halfway through each log, not only to secure the logs but also to allow successive layers (courses) on adjacent walls to lie flat against the log below, in spite of the intervening logs on the other wall. Chinking (filling) the gaps between the logs with masonry or loose material is kept to a minimum in round-log houses. In the best examples, a lengthwise groove is cut on the bottom of each log, and after careful use of a scribing tool, the log is fitted to the exact upper contour of the log below. Before setting the log in place, the groove is filled with moss (or nowadays, fiberglass batts) to form a weathertight seal.

Hewn-log construction, with visible chinking between logs, is a purely American style, without direct precedent in Europe, although both hewing and masonry infill have been used for centuries in European timber-frame construction. Chinking is sometimes considered an inferior technique, even though it does allow use of uneven logs, since the width of the filler material can vary with the width of the spaces. Chinked construction uses fewer logs and can also be very attractive. Be aware, though, that good, careful chinking is tedious work and that a poorly chinked house requires constant maintenance.

Hewing logs reduces their weight considerably, an advantage to most owner/builders. The practice also removes much of the less durable sapwood from the logs, lengthening the life span of the house. Beyond these, other reasons for hewing seem to be personal (a preference for flat walls, and the feeling of a more refined appearance compared to round logs). Many hewn-log buildings were originally intended to be sided over with clapboards, and many were, usually after the owner/builder had

Photo 4.1 ***The Langsner homeplace, winter 1980–81. The long eaves wall faces due south. A solar greenhouse will be constructed just beneath the ground floor windows. A porch and balcony are planned for the shaded west side.***

performed more pressing tasks. Siding, of course, requires flat-sided logs and flush corner notches whose ends do not project.

A major consideration in building with logs is that green wood shrinks as it dries. The diameter of a typical green log can shrink about 4 percent (roughly ½ inch per 12 inches of diameter). Lengthwise shrinkage is negligible. To complicate matters, wood is hydroscopic, meaning it takes on or loses water with changes in humidity. Logs continually fluctuate between swelling and shrinking, taking on moisture in humid conditions and losing moisture in dry conditions. Most shrinkage takes place during the first year after cutting. Hewing exposes edge grain and speeds shrinkage; deep checking is more likely to be a problem with hewn logs (especially green logs) than with round logs.

Due to shrinkage and log movement, walls built with green logs must be somewhat higher than needed. More important, since vertical members (such as door and window frames) will not shrink with the horizontal logs, they must either be installed after most shrinkage has taken place, or especially modified by the addition of recessed grooves to accommodate log movement.

In most log structures, walls are constructed at right angles, forming four-sided pens. While it's possible to build other log polygons, none are as strong as a regular rectangle. With square corners, notching consists of cutting 90-degree angles on both sides of the notch. Other wall angles require cutouts that include acute angles, which are much weaker than right angles. Log pens are inevitably weakened by cutouts in the walls, such as door, window and chimney openings. To keep the pen strong, limit the number and size of openings, and place them with concern for the integrity of the structure. Cutouts should be several feet from corners. Be sure to use full-length logs for courses above doorways and windows.

European logbuilders used mortise-and-tenon timber-framing techniques to lock logs in place around all openings. The joinery is complicated by the fact that vertical posts are stable in length, whereas horizontal timbers expand and contract with changes in humidity. In America, door and window openings are often framed with heavy planking spiked or pegged directly into the end grain of the logs. More elaborate systems, which use floating frames to allow for log movement, may not be as strong as the simple spiked method, but the advantage of the floating frames is that enclosed doors and windows are not affected by log movement.

The size of a log pen is determined by the length of available logs. Very long logs are often difficult to notch due to the differences in diameter between the butts and tips of the logs. Weight also limits practical log lengths; green tulip poplar, for instance, weighs about 50 pounds per cubic foot. A cubic foot of wet red oak can weigh over 75 pounds.

Walls can be extended even with short logs by splicing separate log sections together or using short pieces built up around door and window openings. Both these methods do, however, weaken the log pen to some extent.

LOG-HOUSE TYPES

Because building a log house is a very personal endeavor, I suggest early development of the image you want to create, whether it's the Swiss-style "everything in one building" or an American-style log cabin. The Swiss or alpine *Einhaus* (literally, "one house") is more efficient in the long run in materials used, heating requirements, convenience and ease of maintenance, but such large structures are very demanding. Design and execution must be carefully worked out, so the building does not overwhelm the environment and its occupants. The *Einhaus* does not lend itself to modifications at a later date. This is a building for which all details must be carefully considered before construction begins. The

Scandinavian style of building clusters is simpler and can be done over a period of years. The smaller scale results in a more intimate feeling, reminiscent of a tiny village.

MULTIPLE FLOORS? An early decision either to build up (multiple stories), or spread out (single story) will influence many design and construction details. With the exception of some small sheds and very old dwellings, log construction throughout much of Europe is multistoried. The advantages of multiple floors, rather than a rambling, spread-out approach, are numerous. There is a saving in materials and labor in constructing the roof, flooring and foundation. Heating and cooling is simpler and more efficient. The vertical design minimizes interior walls and hallways, resulting in simplified construction and better interior lighting. Dark, cold back rooms can be eliminated. Multiple levels help achieve a sense of privacy in a compact building. Also, building upward increases the view (from upstairs) and uses less ground area. In some cases, the extra height can provide better exposure to sunlight for solar collectors.

On the negative side, multiple stories require stairways, which often take up considerable floor space. Stairs may be a hazard in emergency situations or for handicapped people. Because of the extra weight per square foot, multistoried buildings require stronger foundations. Tall buildings catch more wind than low structures, and greater roof overhangs are needed to protect the walls from rain.

From a builder's viewpoint, multistoried construction can utilize short log lengths, although getting logs up to a second-story wall requires extra effort.

LOG-PEN DESIGNS. The traditional American log house often began as a single pen (or box), commonly measuring 16 by 20 feet. This is a good size for several reasons. A 12-inch by 20-foot hewn log will weigh from 400 to 600 pounds, about the limit that two people can handle without special equipment. The 16-foot dimension also simplifies joist and rafter design because greater spans require larger timbers or supporting girders and roof trusses, complicating construction and wasting valuable interior space.

The basic pen can often be enlarged vertically or horizontally. An open ceiling in a small house results in a feeling of spaciousness, or the same space can be closed in as an attic for storage or as a guest room. A partial loft or balcony is also effective. Roof pitch makes an enormous difference in usable attic space and overall appearance of a building, so consider your need for space when you decide on the roof pitch. A root cellar or basement can also increase available space in a pen design, especially on sloped building sites. For food storage, the floor can be bare earth.

The basic pen can also be expanded horizontally, usually at the gable ends of the original enclosure. Extending the eaves of a single-story building results in a very low addition—good for storage but not for living space. With a full two-story pen, eaves can be extended without blocking upstairs windows. Log extensions should be made during the original construction process; notching into a finished log wall is difficult. In our area, sheds are sometimes extended on both sides of the original pen. One side is often left open as a roofed porch. The front porch overhang protects the entrance, but often results in a dark interior. I generally prefer front doors located on a gable side.

Jettied cantilever construction combines vertical and horizontal expansion without enlarging the ground floor area. Overhanging floors protect lower stories from the weather. This method is common in Europe, especially in densely populated areas where land is very expensive. One traditional way to expand log pens is to construct two reasonably sized pens separated by a hallway, called a dog trot, with a common roof built over both units. The dog trot is often closed in, creating an extra room with minimal work and materials. Dog-trot houses are usually two stories high. The two pens may be different sizes or proportions. They can also be built at one time or in two stages.

ROOF DESIGN

In folk architecture, roof design is often the most important element in the appearance and construction of a building. Three general types of roofing traditionally have been associated with log structures: the flat shed, the gable and the hip. Roof construction consists of two elements: framing and weatherproof covering. Framing members can be logs, poles, timbers or dimension lumber. The covering is often shingles, but many other materials can be used successfully, such as sod, tile, slate and thatch.

The simplest roof, called a shed, consists of a number of rafters running from one wall directly to the opposite wall in a single, sloping plane. Shed roofs are simple to build, and an advantage is that the roof load bears vertically on the walls. Shed roofs don't create thrust; consequently the roof framing can be kept simple since elaborate bracing for the roof and walls is unnecessary. However, shed construction is seldom used for the main roof of a log structure. Traditional roof coverings, such as shingles, require a fairly steep pitch, or slope, and to use a shed roof with a log pen requires extending one wall vertically. This, of course, is difficult to do since there are no adjacent logs into which to notch the extension wall logs, so conventional framing must be resorted to in most cases, which is an unfortunate compromise to have to make in view of the building's overall traditional design. More basic than this is that simple shed roofing lacks the strength provided by triangulation, a natural feature of gable roofs, which have a peak in the center and incorporate horizontal collar ties to

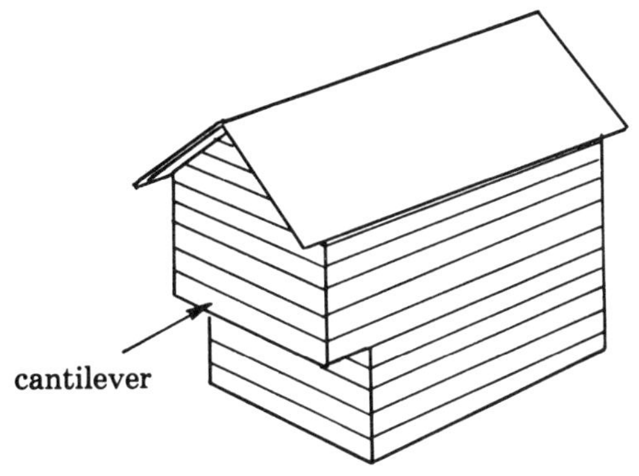

Illus. 4.1. *Jettied construction expands upstairs floor space, while protecting lower floors from inclement weather.*

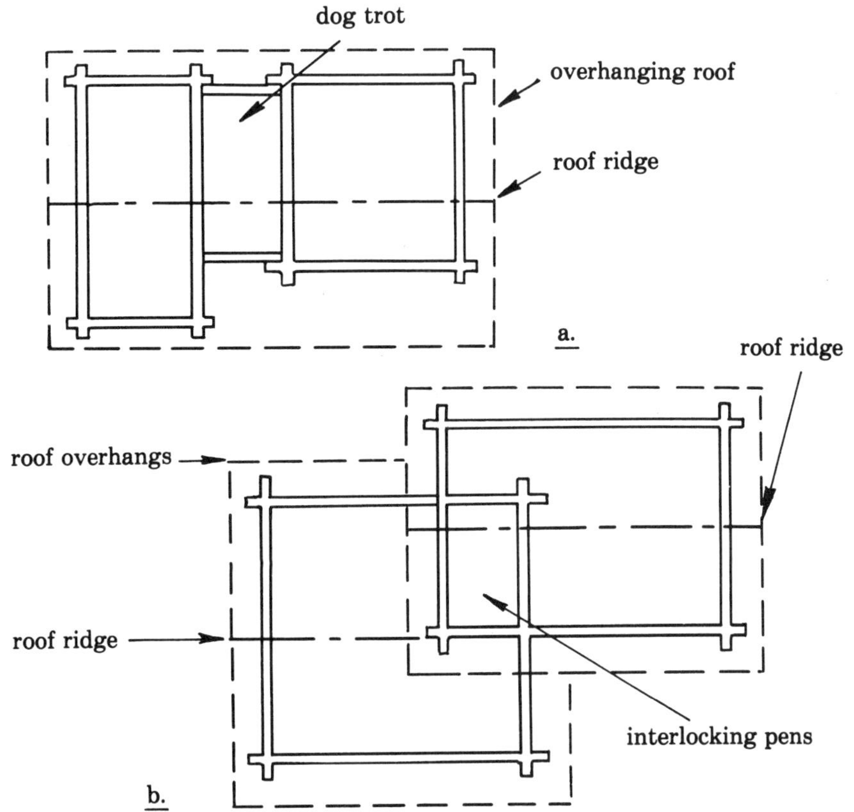

Illus. 4.2. *(a) The dog trot joins two simple log pens under a common roof. The dog trot can be walled in or left open. (b) Interlocked pens used on a split-level site.*

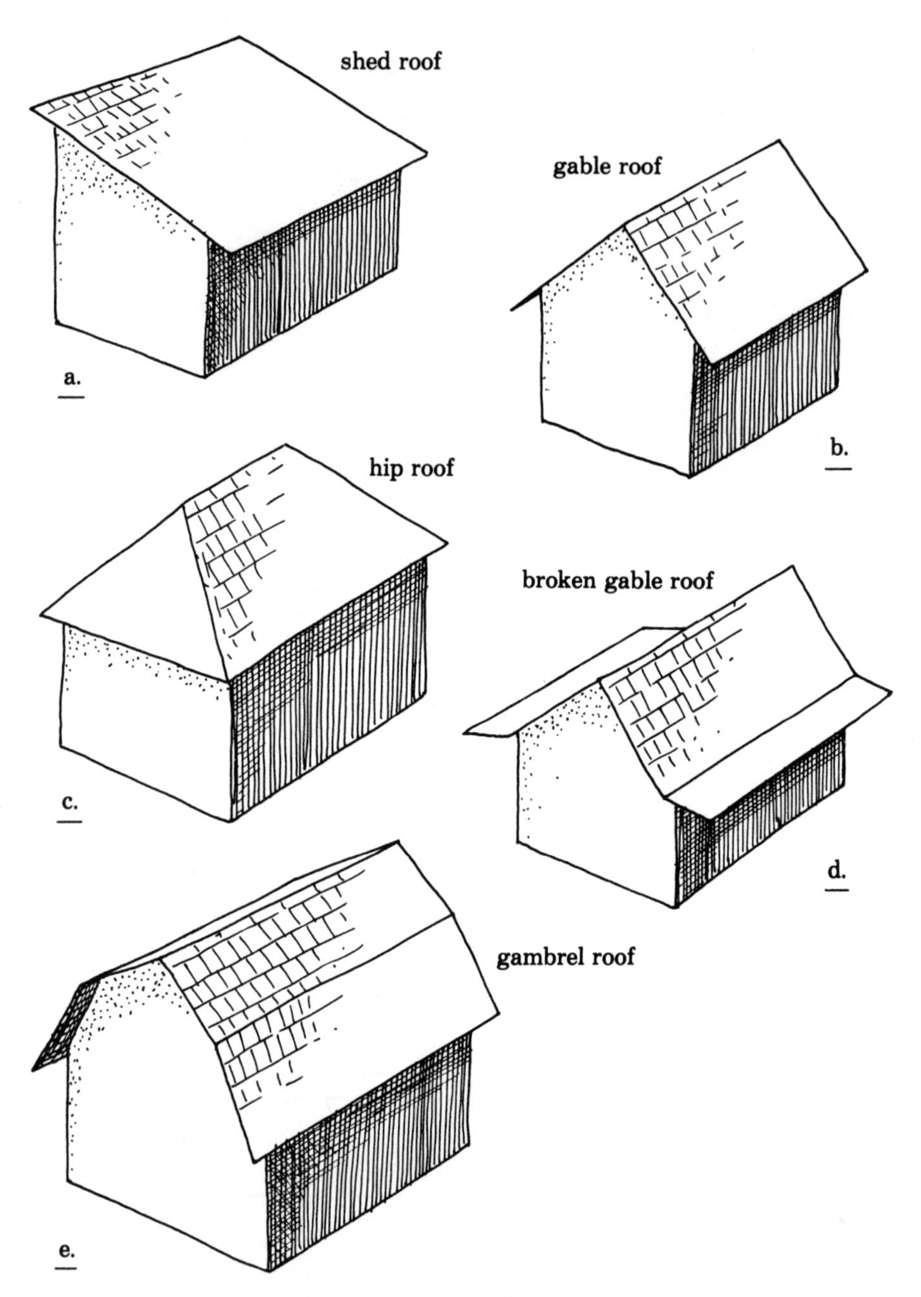

Illus. 4.3. *(a) Shed roof, though simplest to make, is not as sturdy as peaked roof, nor does it permit optimal use of attic space. (b) Classic gable roof. Pitch can vary widely. (c) Hip roof is very strong. Four eaves on roof offer more wall projections than twin-eaved gable. (d) Broken gable increases length of roof overhangs, providing additional protection for walls. (e) Gambrel roof offers the greatest amount of attic space.*

give rigidity to the structure. The long and inherently heavier timber lengths required for a shed roof as compared to a gable roof only compound the problem, and any addition of an elaborate truss system to provide support only cancels the simplicity of the original geometry. Shed roofs generally are successful only in secondary circumstances, notably on so-called shed additions.

A gable roof consists of two sloping planes meeting at a horizontal ridge, like two sides of a triangle. At each end of the building, vertical, triangular extension walls rise from the main walls to the roof peak, filling in the open areas. These triangular extension walls, by the way, are called gables and it is from these that the name of the roof style derives. The advantages of gable construction are considerable, and there are many variations. Gable roofs can be built at any pitch. The space underneath can be used, and the gable-end walls are ideally situated for windows or vents. There are two distinct methods used for framing gable roofs, purlin construction and rafter construction. Sometimes both methods are combined in a single framework.

Purlins are horizontal roof beams that run between gables, supporting the rafters. In purlin log construction, the gables are usually built with logs cut progressively shorter to form a triangular continuation of the end walls. Saddle notches are cut on the undersides of both the gable logs and the purlins. First, a gable log is laid down at each end, then a purlin on each side, running the full length of the building. As the gable logs become smaller, the purlins get closer, forming the roof pitch. An advantage of purlin construction is that the weight of the roof is borne vertically; the roof presses straight down on the gable ends. There is little outward thrust, which is the main weakness of rafter construction. On the other hand, purlin construction tends to be very heavy, and purlin span is limited by the load that a given beam can carry. Trusses over load-bearing interior walls or girders can be built to support longer spans, but, of course, this adds to the weight and complexity of the structure.

In the old days, purlins were often covered with very long shingles called boards in some areas, shakes in others. Now dimension-lumber framing is usually nailed on edge across the purlins, with light horizontal lath then nailed on top, spaced to accommodate the final covering material, usually shingles of some sort, which overlap.

A framed gable roof is built of pairs of sloping rafters that rest on the side walls (not on the gables, as in purlin roofs). At the roof peak, each rafter may be joined to its mate, or to a horizontal ridge instead. Traditionally, horizontal lath is then nailed to the rafters to receive shingles or other covering. Rafters can be straight, round poles, hewn timbers or dimension lumber. Because rafters can be cut to any pitch, the builder can easily adjust dimensions such as attic clearance and eaves projection.

The main consideration in rafter construction is that opposite rafters be held together with horizontal collar ties to counteract the

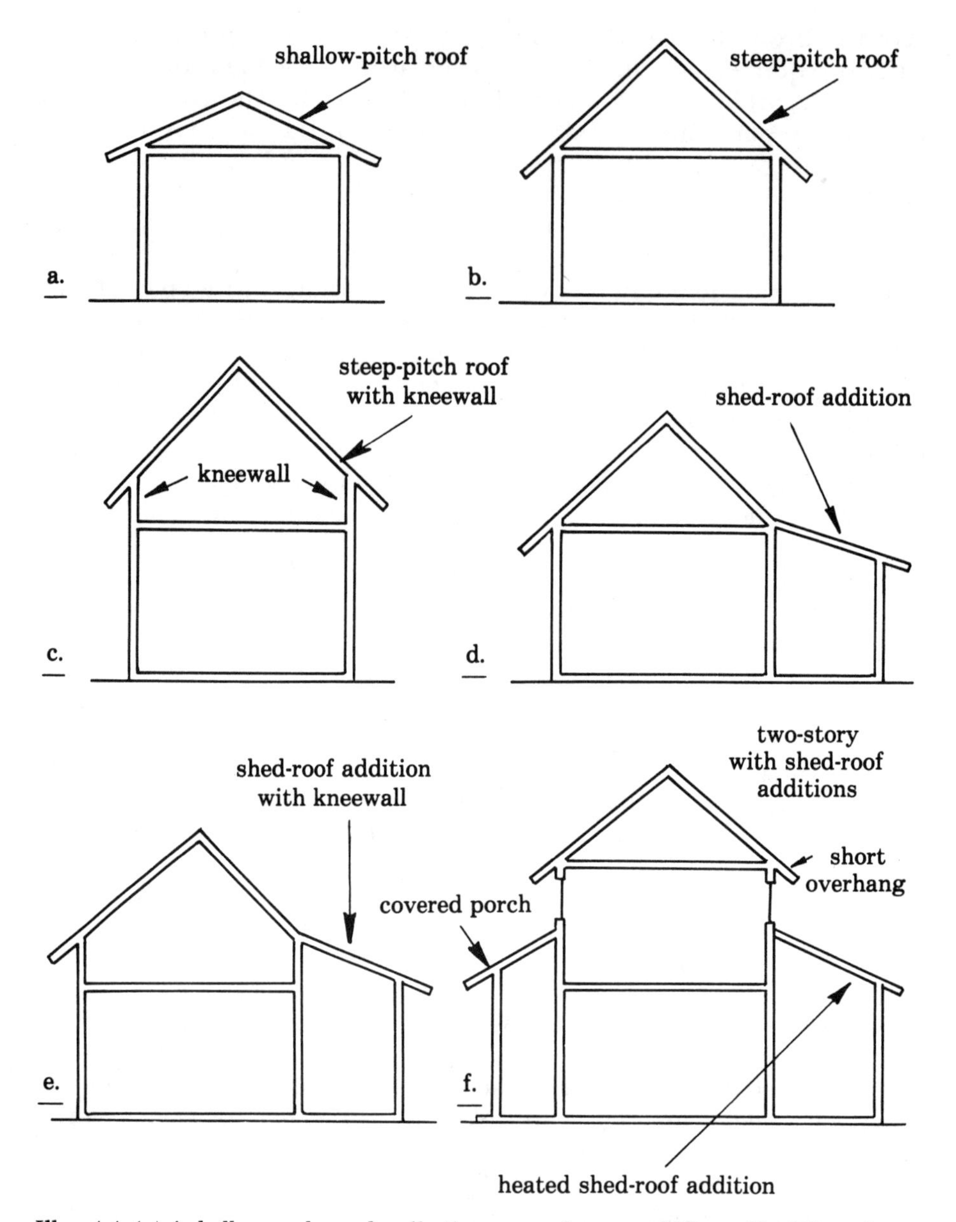

Illus. 4.4. *(a) A shallow angle can be effective on very large, multifloored buildings, but on small, low structures, a shallow pitch wastes potential attic space. (b) Storage attic or sleeping loft created by a steep roof pitch. (c) Two or three extra courses forming a kneewall create upstairs rooms. Also, roof overhang does not block ground floor windows. (d) A shed-roof addition allows expansion of the existing structure. (e) An efficient plan for maximizing living space. (f) Shed additions allow shorter overhangs on the multifloor structure.*

outward thrust caused by the peaked roof bearing down at an angle upon the side walls. Inadequately tied rafters can cause the roof line to sag or the side walls to bulge outward. A ridge beam supported by vertical posts, in addition to collar ties, will restrain the tendency of the roof to spread and sag. Diagonal bracing, nailed from the plate logs at the middle of each side wall to the peaks of the rafters at each gable end wall, will prevent racking (twisting) and add support to the ridge line.

The third traditional roofing style is called a hip. It is similar to a common double-pitch roof such as the gable style, but the top of the gable ends also slope inward, forming four pitches, one above each wall. The shape is a pyramid with two sides pulled back. With its triangulated framework, the hip roof is intrinsically strong, and it is stable against winds from any direction. However, a hip roof with large overhangs results in a dark interior structure. Attic area is also reduced by the sloping end walls. For the attic space to be useful, skylights or dormers are often required. The absence of vertical gable end walls also tends to limit potential flue locations.

Gable and hip roofs can be modified in many interesting and attractive ways, though such alternatives complicate construction. Attic dormers, for instance, are a handsome and practical addition that can be used with any roof plan.

One simple style modification of a gable roof, the broken gable, consists of changing the pitch from a steep to a shallow slope where the roof overhangs the walls. This maintains the interior volume of the attic but increases the height of the roof overhangs. The shallow, overhanging rafters can be supported by knee braces against the walls or with vertical posts to the ground. This variation is particularly effective on small structures.

A gambrel roof is pitched at the ridge like a gable roof, but changes pitch, becoming steeper near the walls. The result is a double slope on each side of the ridge and vertical gable end walls which causes a significant increase in interior volume over the simple gable roof. A gambrel roof is supported by a series of trusses, so the resulting loft area is free of posts or other obstructions. The effect is impressive on large houses and barns, but often rather sad on small structures. A third shallow slope for the eaves is common, and the Swiss sometimes add a small hip at the peak of a gambrel.

Roof pitch is usually described as a ratio between the vertical distance (rise) and the horizontal distance (run). A 4-on-12 pitch, for instance, means that the roof angle rises 4 inches in height for every 12 inches of its horizontal length. The choice of a roof covering is influenced by pitch to a great extent. For wood shingles, a steep pitch of 8 on 12 is probably the minimum, unless layers of roofing paper are installed between each course. A pitch of 8 on 12 is about the steepest a person can work on without special equipment. (With greasy, freshly treated shingles,

Photo 4.2 ***Purlin roof construction covered with straw thatch. A farm shed on a windy subarctic plain; Vääsanlaani, Finland.***

the safe slope is much less.) Where rough-split shingles are used, a 12-on-12 pitch—a perfect 45-degree angle—is not uncommon. A factor to consider is the possibility of installing solar collectors on a south face of the roof. The optimum angle for collectors is the sum of the latitude of your particular location plus 10 degrees. A variation of 10 degrees plus or minus, however, will still result in performance that is 95 percent of what you would get with an ideal pitch.

A steep pitch allows greater space and headroom in attics or upstairs rooms with open ceilings and increases the speed at which rainwater drains off the roof. Steeply pitched roofs are also able to carry greater snow loads than shallow ones. But steep roofs are considerably harder to work on and their increased surface area requires support against wind forces.

A note about roof overhangs: To keep rain away from the side walls they should project 18 to 24 inches on one-story structures, and 24 to 36 inches on buildings with two stories. Overhangs at gable ends, however, can project as far as desired.

FOUNDATIONS

The primary function of a foundation is to support and distribute total building weight. Structures lacking an adequate foundation tend to tip or sink over a period of time. The foundation for a log building should serve other purposes as well. Logs must be protected from boring insects and moisture (which leads to rot caused by fungi). The main sources of moisture are poor drainage, splashing rainwater and moisture in the ground. Continuous-wall foundations also play an important role in retaining household warmth during cold weather.

Many foundation options are available to contemporary logbuilders. Piers or continuous-wall foundations can be constructed with stone-masonry, poured concrete or cast blocks, and various combinations of these materials can be used. Log buildings can also be constructed on steel-reinforced concrete slabs.

Pier foundations are inexpensive and easier to build than continuous foundations, but during cold weather, open piers allow frigid air

Photo 4.3 ***A dry-wall masonry foundation at the Vance Birthplace.***

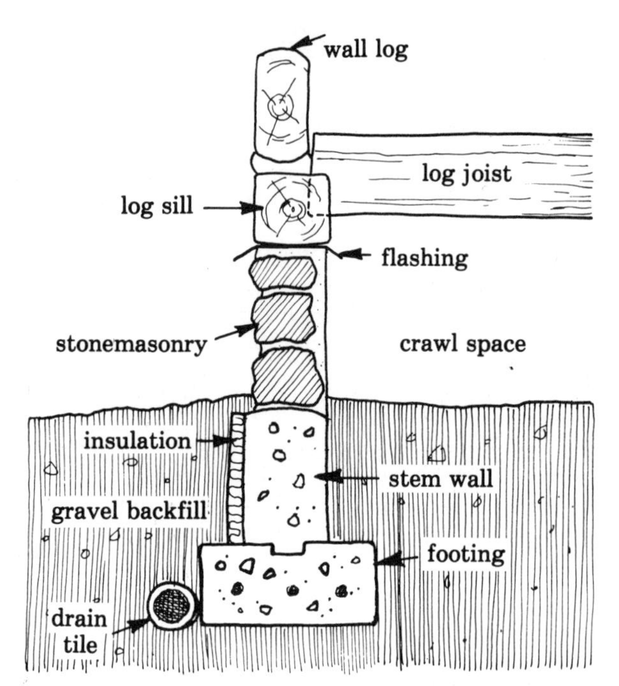

Illus. 4.5. ***A good foundation must have a footing, good drainage and solid soil to rest on. Place insulation outside the foundation for greatest efficiency.***

to blow under flooring, increasing heat loss. Exposed plumbing may freeze with this type of foundation. A continuous insulated skirt should be built between piers to reduce heat loss, and the floors should be heavily insulated.

A continuous-wall foundation creates a draft-free space under the building that tends to stay warm in cold weather and cool in hot weather. Vents are required to allow moisture to escape in warm weather. A continuous wall constructed with native stone is perhaps the most handsome foundation for a log building. The heavy logwork looks strongly supported and the building appears to grow out of the earth, blending harmoniously with its natural surroundings. The cost of materials is low, but good stonework is challenging and time-consuming.

Stonemasonry (stones laid in mortar) is very handsome, but heat loss is high. Poured concrete or blocks, on the other hand, can be insulated on the exterior and then plastered. This option is particularly attractive on hillside sites that would otherwise require the ground floor to be built well above the sloped grade line. Building partially into the earth further conserves heat and helps with summer cooling. In addition, fewer logs are required.

Poured concrete requires carefully constructed forms and large quantities of cement. An alternative is concrete block. New mortarless methods of block construction use fiberglass filaments in a cement base troweled onto the block sides, producing a solid foundation with a stuccolike finish.

SOLAR TEMPERING

A well-insulated foundation and excellent weather sealing are essential to energy-conscious design, but while you are thinking about energy, consider incorporating features that will gain energy for your home, not merely reduce loss. Solar tempering may seem too modern a concept for a rustic log home, but solar design has been a building concept for thousands of years. Here are a number of ways to take advantage of the sun's warmth without making your home look like some futuristic thermal device:

- Buildings should be located to take maximum advantage of winter sun between 9 A.M. and 3 P.M. Place the building in the northern portion of your site to minimize shadows cast over other outdoor areas during sunny weather.
- Properly located windows are the best source of direct-gain solar heating. The optimum orientation is as close to due south as possible. In all northern latitudes (32° to 56°), the south side of a building receives nearly three times as much solar radiation in the winter as the east and west sides. Interestingly, southern orientation is also superior for summer cooling because the higher arc of the midday sun then passes over the roof, reducing direct exposure of the window to the high-intensity sunlight at that time of the day. For maximum heat gain, the width of the south-facing rooms should not exceed 2½ times the height of the south-facing windows (measured from the floor).
- During winter, the north wall of a building receives no direct sunlight. To minimize the shaded area, the north side can be sloped toward the ground. The north wall is a good location for a shed-roofed addition that could serve as a protected entry or storage area. Building into a south-facing hillside is another energy-efficient option.
- Locate rooms that need quick morning warmth to the southeast. An advantage of two-story buildings is that a bedroom and the kitchen can share this exposure, with the upstairs bedroom receiving sunshine first. Daytime work areas should face south. In our case, the workshop, which is used mostly in the afternoon, was an excellent choice for southwestern exposure. Corridors and storage areas should be on the north side.

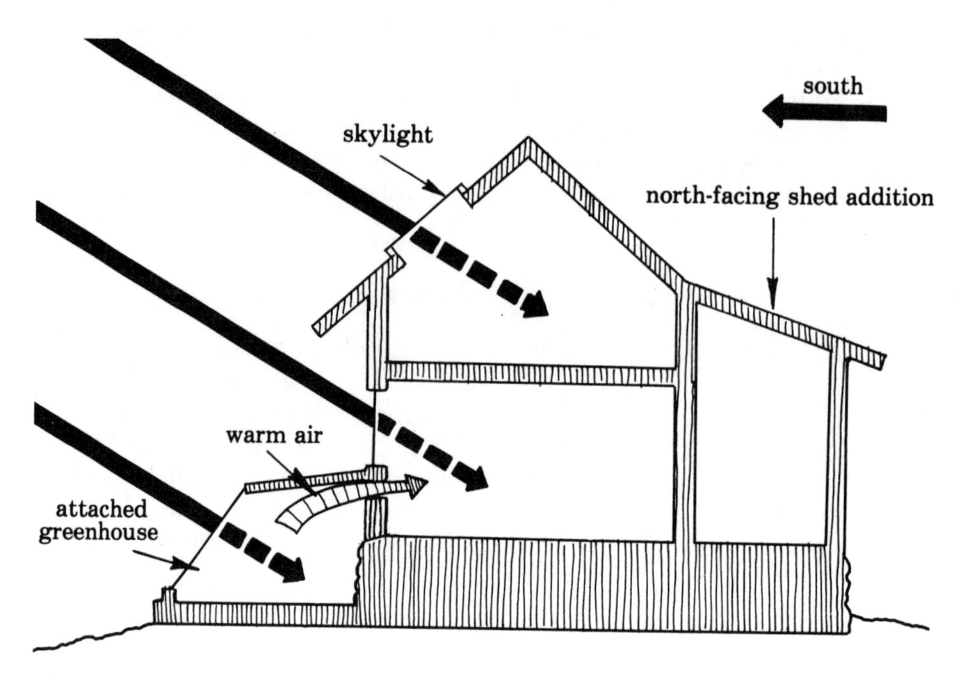

Illus. 4.6. ***South-facing windows, skylights and solar greenhouse provide solar heat, but the glass must be insulated at night, or the heat will escape. North shed could be used for air lock entry, pantry, laundry or storage.***

• Having an entrance with a small room between two doors serves to minimize loss of heated air between living quarters and the outdoors. This vestibule can also serve as a convenient area for removing boots, storing raincoats, hats and mittens. The entry should be located away from prevailing winds.

• As the best source of direct solar gain, the majority of a building's windows should be on the south side, oriented within 25 degrees of due south. You'll still get at least 90 percent of gain within that range. Windows are a major cause of heat loss. North windows should be small and double-glazed to minimize the loss. When the outside temperature is 30°F, and the inside temperature is 68°F, 1 square foot of a single-glazed window loses 20 times more heat than 1 square foot of wall with 3½ inches of fiberglass insulation. The heat loss ratio is about 10:1 between a 6-inch-thick log wall and a single-glazed window.

• Skylights are an excellent means of distributing sunlight to rooms that are not well lit by south-facing windows. Typical applications include deep rooms, lofts, rooms with open ceilings and attics. Install skylights on

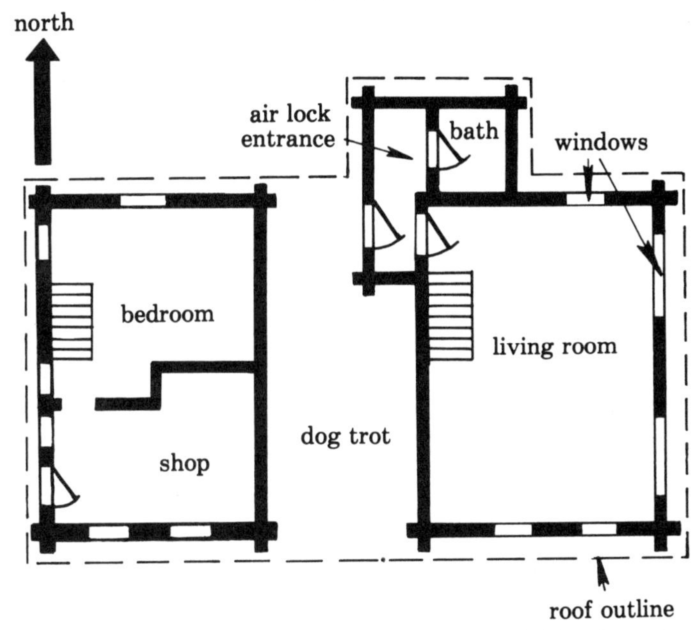

Illus. 4.7. ***Locate rooms to receive most sunlight on the south side. Here, living room gets light from morning till late afternoon; shop remains lit until sundown.***

south-facing roofs to maximize heat gain (north slopes don't receive enough sunlight to provide warmth inside). Double-glazing and movable insulation are recommended.

• Movable insulation refers to shades or panels used to block the loss of heat through windows at night. Night heat loss can be minimized by two-thirds with properly designed insulation. Good movable insulation is much more effective than storm windows or even triple-glazing. In order for a window to receive more heat from the sun than it loses, some nighttime insulation is a must.

• A solar greenhouse attached to the south side of a building can take advantage of surplus energy (not needed by growing plants) as a heat source for adjacent living quarters. Because the intensity of sunlight varies, a heat storage system is required to store and gradually reradiate heat collected during sunny weather. Thermal mass can be a masonry floor or wall, or a bank of dark-colored containers (such as oil drums) filled with water. For maximum efficiency, a solar greenhouse utilizes double-glazing and movable window insulation. In simple systems, warm air moves to living quarters through vents in the common wall.

• Shading devices minimize the solar energy striking south-facing windows during the summer months. The shade can be a solid projection located above the top of a window or an overhanging eave. The horizontal projection in southern latitudes (36°SL) should be one-quarter the height of the window opening and in northern latitudes (48°NL) one-half the height.

Vines growing on a trellis make good shades because they bear and shed leaves with the seasons. Overhangs are inadequate for west side shading because of the low sun angle during the afternoon. Carefully located trees and tall shrubs can also be effective in blocking unwanted solar gain. (Deciduous trees drop their leaves in fall, allowing sunlight to strike the building during winter.) Interior drapes and shades can also be used to block out summer sun.

INSULATION

Insulation not only minimizes heat loss in winter, it also makes the living space more comfortable. The first rule in dealing with insulation is that heat always seeks cold. Insulation resists heat transfer by slowing the movement of heat from warm to cold areas while keeping cold outside. In warm weather, insulation helps keep heat out during the day, especially if you have masonry areas to serve as a heat sink for excess heat.

The insulative quality of a material is measured as R-value—the resistance to heat flow of the material. R-values are directly proportional: A material with an R-value of R-2 resists heat half as well as an R-4 material. R-values are compounded by simple addition. Two inches of a material rated R-4 per inch equal R-8.

Occasionally insulation is confused with thermal mass. Thermal mass holds heat and reradiates it slowly when room temperature falls below that of the mass. A masonry oven is an excellent example of thermal mass because it absorbs, holds and radiates heat well. Masonry is a poor insulator but it stores a lot of heat. Insulation materials, such as fiberglass, resist heat flow but store little heat for future use. If you pick up a cold axe, the handle feels warmer than the axe head. Actually, both are the same temperature. The axe head seems colder because it absorbs heat more readily from your fingers. In contrast, the higher insulative quality of the wood handle resists transfer of heat. To be effective for heat storage, a substance must have a fairly high heat storage capacity—and good conductivity, so the heat can get in and out.

Thermal mass used to store solar heat should be enclosed behind glass to prevent heat from radiating back outside too quickly. Glass allows sunlight to pass, but blocks lower-frequency heat radiation. Masonry behind glass heats up when it's sunny, and the heat is trapped by the glass. When the sun goes down, the heat radiates to the room where the masonry is located.

Most wood species used for log walls are rated about R-1 per inch (see table 4.1), or R-6 for a 6-inch wall, not good compared to fiberglass, but better than a stone wall. The highest-rated commonly used wood is white pine, which averages R-1.3 per inch, or about R-8 for a 6-inch thickness. Wood does increase slightly in R-value as the surrounding temperature falls. For instance, white pine rated R-1.3 at 73°F averages R-1.8 at 41°F (this is the mean temperature; the cold side was 14°F, and the warm side was 68°F). For paneling or flooring this difference would not be significant, but it is good news for log houses with walls 6 to 8 inches thick.

With hewn logs, the R-value of the chinking should be considered. Traditional chinking, consisting of wood sticks and clay, has an R-value considerably lower than the log walls. We chinked with stucco on both sides of a 3-inch band of foam insulation. The chinking R-value is about R-12, improving the value for the wall.

The low R-value of log walls can be compensated for with careful design and construction. By incorporating passive solar design with proper weather stripping and insulation, one can build an energy-conserving log house. In addition, heating with a woodstove or furnace, rather than a low-efficiency fireplace, can keep your fuel requirements well within bounds.

The main areas of heat loss in most houses are the roof (or upstairs ceiling) and windows. The cracks and openings of a poorly sealed house are the third main element in heat loss. A commonly used rule of thumb for ceiling or roofing insulation is to insulate at twice the R-value of the walls. For log construction, the range would be R-12 to R-24. The ground floor is generally insulated at the R-value of the walls. Movable insulation (thermal shutters or panels) is highly recommended for all windows, solar-tempered or not. Exterior doors can have insulation in a hollow core. Log checks and cracks should be filled with insulation and sealed with dark-colored caulking. Log houses with open-beam ceilings can be insulated between the beams with wallboard over the insulation.

The most common insulation materials used by logbuilders are foam panels and fiberglass blankets. Polyethylene foam tubing is very useful for insulating gaps and crevices.

Styrofoam is sold in 2 by 8-foot and 4 by 8-foot panels in thicknesses of 1, 1½ and 2 inches. It has a characteristic blue color, an R value of 5.5 per inch and a moisture-resistant closed-cell structure. When Styrofoam is used as the primary insulation no vapor barrier is required between ceiling (or flooring), sheathing and the insulation. White polystyrene (beadboard) is less durable, not moisture resistant, and rated R-3.5 per inch. Local manufacturers often sell beadboard for less than half the cost of blue foam, but you're usually better off with Styrofoam, except in cases where you may not want a vapor barrier. Both of these materials are flammable and release toxic gases when burning. Foam insulation should be sheathed with fire-resistant materials, such as gypsum board, to minimize available air in case of fire.

Table 4.1 **R-Value of Trees Used in Log Buildings**

Tree Type	R-Value per Inch
Northern white cedar	1.41
White pine	1.31
Balsam fir	1.27
Yellow poplar	1.13
Western red cedar	1.09
Redwood	1.00
Douglas fir	0.99
Red oak	0.79

Fiberglass insulation is generally sold in rolls with a foil or paper backing. The foil offers some moisture resistance, but should be augmented with polyethylene sheets. The most common thicknesses are 3½ and 5½ inches, in widths that fit between framing spaced either 16 or 24 inches apart. Fiberglass insulation is fire resistant and rated at R-3.5 per inch. In Switzerland, I learned to cut fiberglass insulation by chopping it with an axe. Use a scrap plank as a chopping board beneath the blanket being cut. This reduces handling, which can cause irritation. In fact, gloves, goggles and a face mask are helpful when you work with it.

UTILITIES

Log walls do not provide a built-in cavity for wires or plumbing, but electrical wiring or plumbing can pass through log walls if careful planning is done in advance. Construction is slower and more complicated, and future access for repairs or modification is difficult. A great deal of labor can be saved by casting passageways for water and drainage systems through concrete foundation walls. A short length of 4-inch corrugated plastic drainpipe makes an excellent form.

Most plumbing can be concealed in crawl spaces or behind counters. Copper pipe can be used for exposed runs, such as hot water heater connections between a woodstove and a storage tank. Neatly installed copper pipes polished with steel wool and varnished turn a potential eyesore into an attractive feature. An additional advantage of external plumbing is the fact that repairs and modifications are easy to carry out.

Electrical wiring can be routed through the attic and crawl spaces, beneath moldings, behind furniture and along unseen sides of door and window framing. Inconspicuous brown metal coverings are available for exposed runs. The chinking strip between hewn logs is another location for wires or outlet boxes. Electrical wiring should be isolated in metal conduits, away from combustible insulation material.

Where floor or ceiling joists block straight runs of pipe or wiring, you can bore a hole through the middle of a joist with negligible structural damage. However, holes or notches cut through the top or bottom of a beam greatly diminish overall strength and should always be avoided.

HEATING SYSTEMS

Houses with an open floor plan (minimal dividing walls) are often easier to heat than those with enclosed hallways and numerous rooms, especially with a woodstove placed centrally. Woodburners and solar panels can heat water. Supplementary heating units, such as the new generation of efficient kerosene or quartz space heaters, can be used during extremely harsh weather, or in rooms that require occasional heating.

We use wood heat, because on our property many more dead trees rot every year than we could possibly use for fuel. Chips from hewing logs provided a good part of our heating and cooking fuel for about two years. The common complaint that wood heat is the cause of extremely dry air is not accurate. Dry household air in winter is the result of air infiltration—leaky house construction. Cool air holds less moisture than warm air. Cold, dry outside air enters the house and picks up interior moisture as the air warms up. The warm, moist air then leaks out of the house to be exchanged with more cold, dry air. Most woodburning systems do contribute to drafts, increasing this drying cycle, but some old woodstoves were manufactured with a separate air source (ducts from outside the house to the burner) and you can easily rig vents to provide your woodburner with outside air. This will avoid extreme dryness and slow down heat loss caused by warm room air being sucked up the chimney.

Many American logbuilders consider a stone fireplace to be a necessity, almost a defining factor of a log house. While I admire the beauty of a carefully made, open fireplace, I feel that the negative aspects outweigh the aesthetic pleasures. Fireplaces are inefficient fuel consumers; they waste space, cause drafts, tend to be dirty and unsafe, and are difficult to construct. If you have to have one, at least use it with a high-efficiency woodstove insert.

Fireplace or woodstove flues can be built within a house or onto the exterior. The advantage of interior placement is that the flue then contributes to indoor heating rather than simply radiating heat to the outside. Since the flue stays warm, creosote formation (caused by condensation in a cold flue) is also minimized. Fireplaces and masonry flues should be built over a large, solid footing to distribute their enormous weight.

Any heater requires a safe, efficient flue to carry smoke outside the house. The basic choice is masonry or metal. Exterior masonry flues can be insulated to minimize creosote buildup (a major cause of wood-heat fires).

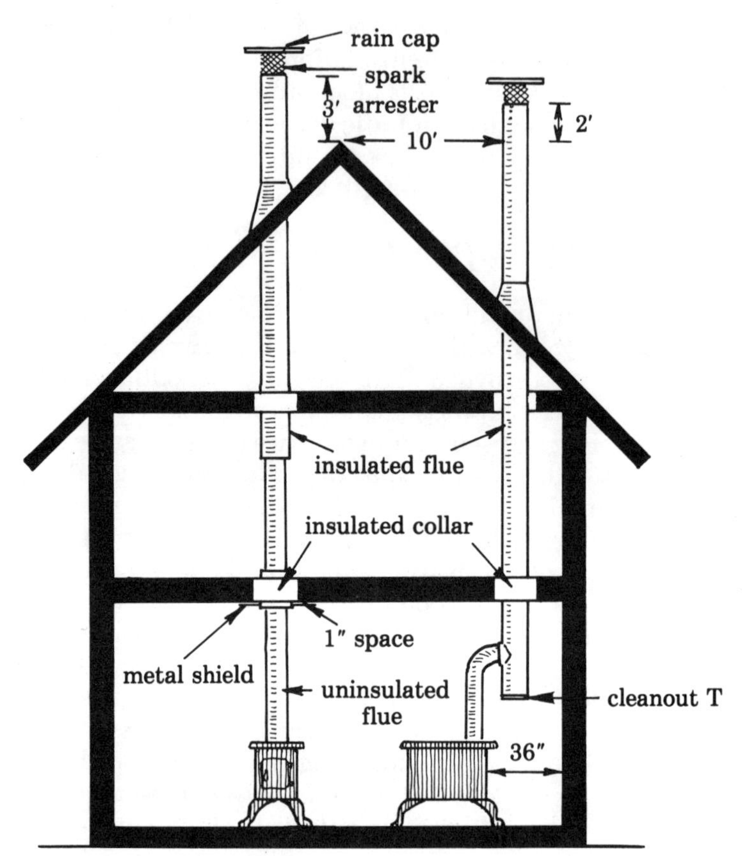

Illus. 4.8. ***Examples of safely installed stoves. Pipe sections are screwed or riveted together. Double-wall insulated pipe is used to prevent fires where the flue passes through the floors, and on the exterior of the house to prevent creosote buildup. A cleanout T makes periodic removal of creosote easier. A spark arrester is a good idea on a wooden shingle roof, and the rain cap keeps water out of the flue. If the flue is located near the roof ridge, it must extend 3 feet above the ridge. You can reduce the distance to 2 feet if the flue is more than 10 feet from the ridge. Keep the stove at least 36 inches from combustible surfaces.***

Most masonry flues consist of special ceramic liners faced with brick, special cement blocks, or stone. Interior cavities can be filled with a fireproof insulation, such as vermiculite.

Ordinary sheet metal stovepipe may be used as a connector between stoves and permanent flues. Never use stovepipe for permanent flues; specially constructed, double-wall, insulated stainless steel chimney pipe is now available. These chimneys, rated for all fuels, are as safe as masonry flues with ceramic liners. Insulated chimney pipes can be attached to ceiling joists, to a wall bracket, or suspended from roof rafters with a wide range of commercially available supports. The exterior diameter

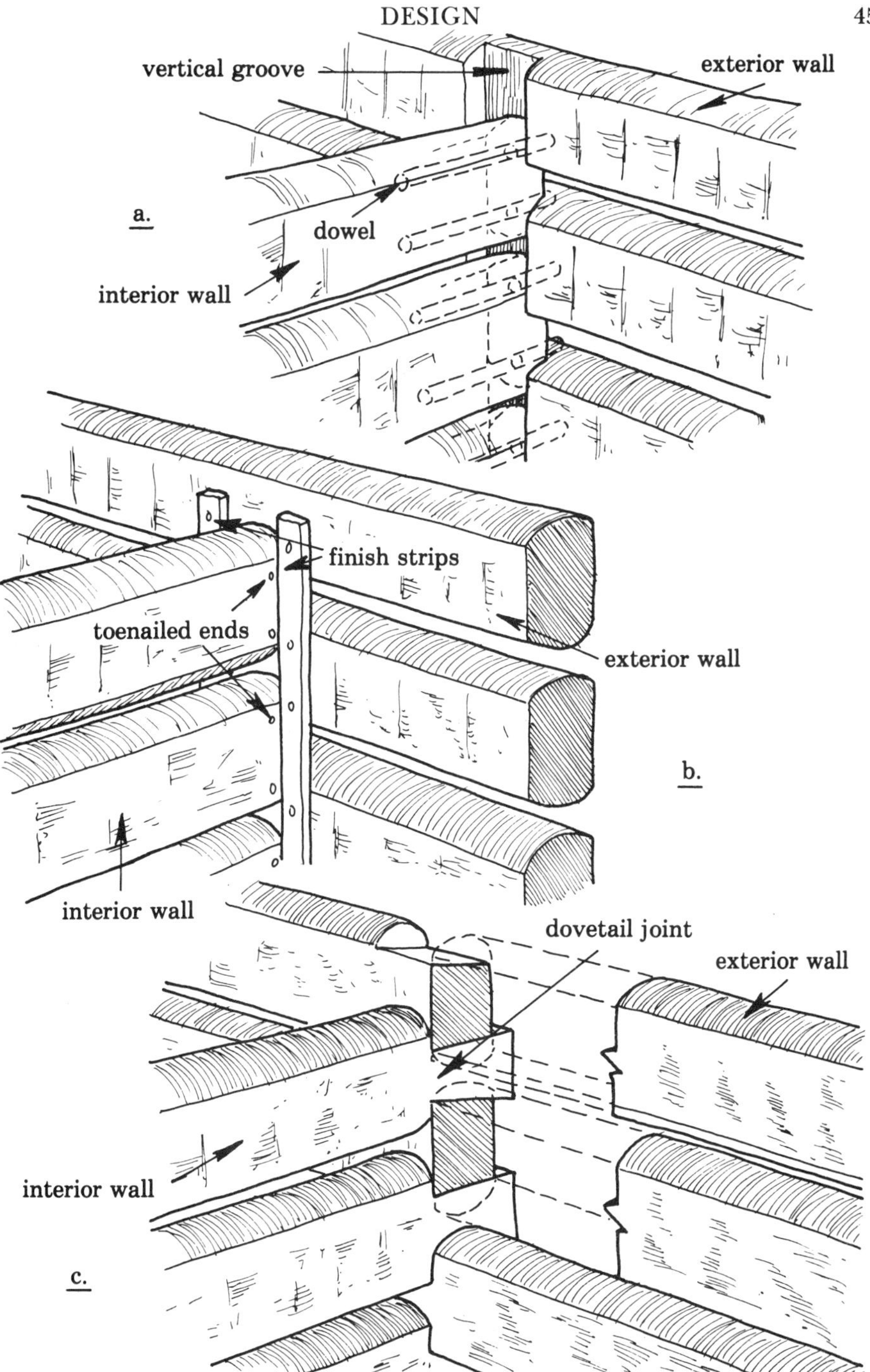

Illus. 4.9. *Three ways to join intersecting walls. (a) An intersecting wall set in a vertical groove and fastened with dowels is sturdy and attractive. (b) Raised-channel method is easiest: face-nail a wood strip on each side of intersecting wall to form a channel. For additional support, toenail interior wall logs to exterior wall. (c) Notched dovetail interior walls must be built at the same time as exterior walls.*

of the 6-inch flue is only 8 inches, and since a mere 2-inch clearance from combustible material is safe, it is easy to install metal-insulated flues unobtrusively.

The general safety rule for flue height is that the open end must be at least 3 feet above the roof penetration, or 2 feet higher than any part of the roof within a 10-foot horizontal radius. Locating a flue at or near a roof ridge simplifies construction and maintenance. Straddling a ridge is easier than working on a steep pitch. Chimneys that extend over 4 feet above the roof surface require steel bracing and may be hard to clean. All chimneys should be provided with rain caps and spark arresters, which also serve as bird screens. Locate smoke detectors and fire extinguishers on every floor, including the attic.

INTERSECTING WALLS

Intersecting walls can be built during original construction, or added later. The best construction method is to notch intersecting wall logs into the basic pen as the building goes up. Fully integrated intersecting walls brace the overall structure, reinforce the basic pen and support the weight of upstairs flooring and purlin roofing. A major advantage also is that as integrated log walls season, they shrink and expand along with the other logs. If you choose to add intersecting log walls later on, you will have to cut shallow grooves in the original wall to receive the log ends of the added wall or create a channel with wood strips on each side of the wall end (see illus. 4.9). With chinked logs, you should also use small blocks between each course as spacers. Toenail the new logs to the original wall with 8-inch spikes or secure the logs with 1-inch-diameter hardwood dowels inserted from the exterior side of the original wall into the end grain of the logs being added. Also incorporate a few heavy bolted angle irons between the original pen and the added wall.

Intersecting walls can also be framed with paneled dimension lumber or large timbers left exposed, with stucco between them. Gypsum wallboard provides an opportunity to incorporate light colors as a nice contrast to the overall "woody" quality of a log house, but the major problem with framed walls is that log shrinkage can result in buckled vertical framing, as well as arched joists and flooring if the framing is tied directly to the logs. Instead, anchor the framed walls to the floor only, floating the wall in a groove at the walls and ceiling. The channel can simply be cut into the original logs and joists, or built up by nailing spaced 2 × 4s to the wall and ceiling.

CHAPTER V

LOGS

Logbuilding requires logs, lots of them. Some species of trees are better for logbuilding than others and some will be more available to you than others. Individual specimens of a given species may vary considerably. The quality of the logs will strongly influence the success of the home you build. With any luck, you can build the house with trees from your own woodlot, but it's much better to get quality trees from an outside source than to try to make do with inferior trees from your own lot. You can always supplement your own trees with timber from a neighbor or a local dealer.

Under suitable conditions, a log home is a very durable structure, but logs, like any wood, are subject to insects, fungi and other destructive forces. Understanding the things that damage wood is an important part of building a home that will last. If you decide to do some of your own logging, you'll need the proper equipment and plenty of caution. It's hard, dangerous work, but the rewards are great, both in personal satisfaction and in the benefit of being able to get superior logs with little damage to the forest. There is also the fine feeling of building with materials from your own land.

LOGS FOR BUILDING

Round-log structures require higher-grade materials than hewn-log buildings. Because round logs usually have a lateral groove for a tight fit, they must be cut from straight trees with little taper. Hewn logs can be less even because the chinking easily fills in varying spaces. Wall logs can be anywhere from 8 to 24 inches in diameter, but it's best to keep the logs in any one building close to the same diameter, within 10 inches, for the sake of appearance. Very large logs (18 to 24 inches in diameter) are better suited to hewn work, since hewing reduces their weight and flattens out the logs. Large logs otherwise produce a "corduroy" effect, bulging out from the plane of the wall. The logs for our house varied between 12 and 18 inches in diameter.

What's a good log diameter for building? Large-diameter wall logs look impressive, but they are also heavy. With hewn construction, the waste is proportionately larger as diameter is increased. Small-diameter logs can look good, and they are easy to handle. But a major disadvantage of thin logs is that because you'll need more logs to make a wall of a given height, more notches and more bands of chinking are required than for thicker logs, increasing the time required to raise the wall. Also, small logs tend to be mostly sapwood (which deteriorates faster than heartwood). Overall, 10- to 16-inch logs are a good range to work with.

Curved and crooked logs can also be used. With careful placement, the visual effect is very appealing. We have several S-curved logs on our house that I am quite fond of. Japanese carpenters are masters at using deformed logs, especially as exposed beams crossing open ceilings or as vertical posts. Working imperfect logs does require special consideration of the nature of the wood and the possible consequences of its use. The curved butt section of a log that grew on a slope may contain reaction wood, irregular (deformed) wood cells that grew to compensate for the lean of the tree. Reaction wood often splits badly. I have seen (and heard) a crack open up as I hewed away at the butt of a tree that grew on a steep slope.

In addition to good wall logs, you'll need special logs for the sills, plates, joists and rafters, and wood for other parts of the house. The specialized logs are generally shorter and thinner than wall logs, but straight, even logs that don't taper drastically are still best. Joists can be round logs, or square-hewn timbers. For round ones, 8 to 10 inches in diameter is enough, or the same diagonal distance for hewn logs. The diameter of logs to be hewn is the distance between opposite corners—the diagonal—of the eventual hewn piece. Slightly waney (rounded) edges are usually acceptable for joists, but knots and defects in joists should be kept to a minimum. Roof rafters can be as small as 5 or 6 inches in diameter. If you have good, straight trees with large-diameter butts, the lower portions can be sawed into flooring, door and window jambs or cabinet wood.

Before you begin selecting logs, decide on one of two types of structure: A traditional American log house made from full-length logs with door, window and fireplace openings cut out after the walls are raised, or a house utilizing short log lengths, with the openings made as part of the construction. The latter method saves a considerable amount of timber and results in better log utilization since odd-length logs can be used. The openings are convenient to go in and out of during construction, and they serve as ledges to stand on while you are working. There are two disadvantages: During assembly the pen broken by half-completed openings is less stable (braces must be nailed to the walls to keep cut-off log ends in proper alignment), and the list of materials is also much more complicated than for a full-length log house.

Think about the timing of your construction schedule, and whether or not you intend to build with green or seasoned logs. North American logbuilding has usually been done with green logs. Seasoning time is avoided, and green logs work very nicely. Also, as logs season after construction, they set. Load-bearing faces of notches actually tighten up. European logbuilders have generally preferred to use seasoned logs. The delay is three to five years, but weight is reduced as much as 50 percent. Seasoned logs are less likely to warp or twist on the building. Controlled seasoning can prevent mold development and checking.

Naturally, the species of trees you select will affect the quality of the home you build, though species selection is often determined more by what is locally available than anything else. Many of the characteristics of good building logs are shared by many conifers and several deciduous species. Prime logs are not only straight with minimal taper, they are also straight-grained, relatively free of knots, and resistant to fungi and insect invasion. Good logs are lightweight, stable during seasoning, and well able to stand the elements. Spiral grain, often indicated by a spiraling bark pattern, may result in the formation of a twist during seasoning and should be avoided.

A wide range of conifers grows in the forests across Canada and the northern United States. In the western states and Canada, white cedar is the logbuilder's prime wood. Western red cedar, Douglas fir and western hemlock are also highly valued. On the Pacific coast, many fine old log houses were built with redwood. Various pines and spruces are used in many buildings. Other species that should not be overlooked include larch and balsam. In the eastern United States, the traditional first choice was chestnut, which is now unavailable due to the chestnut blight that decimated the species throughout the region. Pines are commonly used now. Many log houses have also been built with yellow tulip poplar (ours) and red oak; both hew beautifully when green.

CAUSES OF WOOD DECAY

To keep a log house sturdy, protect the wood from decay fungi, insect attack and weathering. If these challenges are met, log structures can last for many generations. Proof of this is the existence of thousands of centuries-old log and timber-frame structures in Europe. Careful design and construction have kept the wooden houses of Europe serviceable without chemical preservatives. Maintenance must be continuous, but it need not be very time-consuming if a structure is carefully conceived and built.

FUNGI. Surface molds and sap-stain fungi alter the appearance of wood, but do not cause structural damage. Decay fungi actually dissolve the cell wall material by enzyme action. Wood scientists have demonstrated that decay fungi require four specific conditions for growth. Fungal decay can be prevented or controlled by eliminating any one of the following environmental factors, or by applying chemical fungicides.

- Decay fungi thrive at temperatures ranging from 75° to 90°F. Growth stops below 40° and above 105°F.
- The optimum moisture content in wood for fungal growth is 30 percent. Growth stops at 20 percent, which is slightly above the moisture content of most air-dried wood. The water vapor in humid air alone is not

enough to support decay fungi, but will permit development of some harmless mold. Wood will not decay if it is kept air dry, nor will decay already present progress further.

- Fungi require air. If wood is water-soaked, there will not be enough air to support development of decay fungi.
- Decay fungi live on cellulose and lignin. Sapwood is thus much more susceptible to fungal rot than heartwood because of its higher content of these substances. Heartwood resistance to decay varies considerably among species (see table 5.1).

Table 5.1 **Comparative Resistance to Decay of Heartwood of Commonly Used Building Logs**

Very Resistant	Moderately Resistant	Resistant or Slightly Resistant	
Catalpa	Douglas fir	Alder	Hemlock
Cedars	Larch, western	Ash	Hickory
Chestnut	Locust, honey	Aspen	Maple
Locust, black	Oak, swamp chestnut	Basswood	Oak, red and black
Oak, Oregon	Pine, eastern white	Buckeye	Pines, most
Oak, white	Pine, southern	Butternut	Poplar, yellow
Redwood		Cottonwood	Spruces
Walnut, black		Firs	

INSECT DAMAGE. Various boring beetles attack host woods at specific stages of their metamorphosis (at certain seasons of the year) if the moisture content is right for them. Ambrosia beetles bore minute holes into green, summer-cut oak. Their food is a fungus that grows in their own damp tunnels. Round-headed powder-post beetles attack seasoned pine, especially in crawl-space floor joists. It's sometimes possible to hear them ticking away inside the wood. Generally, spring and early summer are the times when insects are most active. Protect the wood from them by keeping it away from the ground.

Most woods are susceptible to termite attack. Subterranean termites live in damp ground, building earth ramps and tunnels to the wood, which they eat. Nonsubterranean termites are harder to detect, since they use wood as their home as well as their food supply; they live in a belt stretching from southern California, along the southern United States, to Virginia.

Carpenter ants can crawl into a house, or be introduced when fuel-wood supplies are either brought inside or stored against exterior wooden walls. They hollow out decayed or naturally soft woods, thriving in damp areas, under porches and in crawl spaces. Log buildings are susceptible.

Most insect damage can be avoided by eliminating the conditions that favor their growth. Insect (and fungal) degradation often begins in substructure areas. Dampness due to condensation can be eliminated by laying a continuous film of polyethylene over the bare earth and ventilating the area. The poly must be covered by a layer of sand or pea gravel to keep it in place. In any case, remove wood waste, such as old concrete forms and building scraps. Termite tunnels can often be blocked by extending foundation sill flashing 1 inch into the crawl-space area. The overhanging edge should be bent downward at about 45 degrees. Inspect crawl spaces frequently. A minimum 18-inch clearance between the ground and joist bottoms will promote air circulation and facilitate the inspections.

Seasonal timing of logging and woodworking operations can also play a substantial role in fungus and insect control. The moisture content of living trees varies throughout the year, and timber that is cut when the sap is down has lower moisture content than at other times of the year. But more important, insects and fungi are much less active from late fall through midwinter, making that the best time to cut timber.

Some logbuilders prefer to leave bark on logs until hewing and notching because the bark prevents drying, keeping the wood green and easily workable. This practice is not recommended if fungi or insects are likely to be a problem. Instead, cutting and peeling should be done during the cold months to permit some drying to take place before warm weather. Although it's harder to peel logs in winter, the added protection against fungi and insects makes the extra labor worthwhile.

Try to do any milling of wood by late winter, so the lumber has a chance to begin drying before warm weather. If possible, air dry lumber (and peeled logs) in a drafty shed where the wood is protected from direct exposure to sun and driving rain. With good conditions, most 1-inch-thick lumber will dry to 20 percent moisture content in two or three months. The same lumber, five years later, will be no drier if it stays in the shed. Some softwoods can be summer air dried in 15 to 30 days. The denser hardwoods, like oak, may take 100 days or more. Whole logs, as noted, take longer, up to five years. Air circulation is essential to proper drying. Seasoning wood should be kept well off the ground and lumber should be carefully stacked on spacing strips, called stickers.

Good design and careful construction can prevent fungal growth after a structure is built. Roofing plays a major role in protecting log walls and millwork from decay. Generous roof overhangs and properly maintained gutters will eliminate wind-blown water from contacting the sides of a building. Shingles should be made from decay-resistant wood, or treated with a water-repellent preservative. Assure good ventilation underneath shingles to prevent moisture from collecting, and flash at roof edges to prevent wind-blown water from working underneath the shingles.

ACQUIRING THE LOGS

I've always wanted to clear a building site and surrounding fields, and use the logs to build a house. This is the classic way to do it—unfortunately a rare one these days. It's an experience that should be cherished by anyone who has the opportunity. It's more likely that you have land with a woodlot that you could log selectively to provide building materials. But in any case, the choice is between tackling a major project yourself or hiring some help. If you don't have a usable woodlot, you may be able to buy standing timber from a neighbor or, possibly, the U.S. Forest Service.

Another option is to buy logs from a sawmill or a log broker, avoiding the difficult and risky task of logging and eliminating the expense of logging equipment. You still have the opportunity to choose and examine exactly what you are getting. Although you will miss the experience of logging, which I consider to be an integral aspect of log-building, there is also a considerable savings in time. Logs bought on the market usually sell at the wholesale commercial price for the particular grade and species of each piece. If you buy logs, arrange to have them delivered at the building site. The seller should cover all handling, including unloading, which can be dangerous.

Purchasing and moving an antique house is another option which appeals to some people. Restoration work is not specifically covered in this book, but many aspects, such as site selection, foundation work, log raising, chinking and shingling are all applicable.

LOGGING

If you decide to do your own logging, the first task will be choosing trees to fell. Unless you're clearing a site or field, plan to log selectively. Regrowth of desirable species requires proper groundcover and a controlled degree of light and shade in the logged area. Your state forester, or the U.S. Soil Conservation Service, will provide free assistance in log selection and reforestation planning.

Logging consists of several distinct jobs, all requiring special skills and tools: felling (cutting down trees), limbing (removing the branches), bucking (cutting logs to length) and skidding (hauling logs to a loading area or work site). Tools can be relatively simple and inexpensive, or complex and extremely costly. Old-fashioned, do-it-yourself logging depends on plenty of log contact, and is therefore a risky and dangerous activity. Large-scale methods are entirely mechanized.

FELLING TOOLS

Historically, the logger's tool was the axe. The ultimate axe is probably the American double bit, developed for use in the Maine woods.

The narrow blade and perfect balance make this a wonderfully efficient tool in the hands of a skilled axeman. During the last 100 years or so, axe felling has become a lost art. Crosscut saws (and later, chain saws) greatly changed the world of logging. Nowadays, loggers generally carry a nondescript single-bit axe along with their other gear. Its main function is driving wedges with the blunt end; the axe blade is used for barking and occasional odd uses that may come up.

Photo 5.1 *The wide, heavy crosscut saw on the left is used for bucking. The saw on the right is used for felling. Its thin, narrow blade is less likely to bind in the saw kerf. Tooth patterns for both types of saw are interchangeable.*

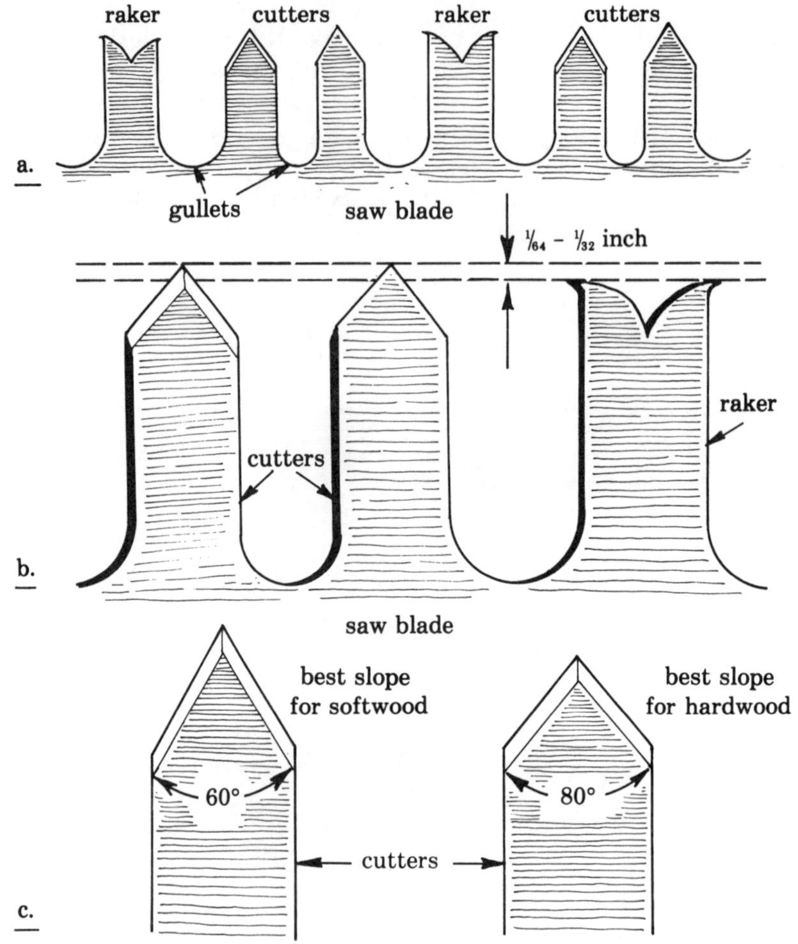

Illus. 5.1. ***Teeth on a crosscut saw. (a) The cutters are set in pairs, with the members of each pair set alternating left and right. The rakers remove chips made by the cutters. (b) Cutters should always be higher than rakers. (c) A narrow angle for the cutters is best for softwood; use a wider angle for hardwood. A good compromise for cutting both is an enclosed angle of 70 degrees.***

CROSSCUT SAW. Don't overlook the potential of the two-man crosscut saw. In its day, it was one of the great inventions of mankind, and it is still an excellent way to cut timber. Cheap, pollution free and dependable, the crosscut requires only the energy provided by you and your sawing partner. Because you use it with someone else, there's a nice social element—and you won't be tempted to defy safety and go logging alone. Crosscut saws are far safer than chain saws.

Crosscut saws have large, intimidating-looking teeth: a row of narrowly spaced cutting teeth and larger raker teeth that scoop out the waste produced by the cutters. The combination of raker and cutting teeth is responsible for fast cutting, and there is less chance of getting hung up on the wood chips. The cutting teeth are angled slightly beyond the plane of the blade to score the wood fibers. Between the cutting teeth and the rakers are large spaces called gullets where the chips collect as the rakers push them along.

Felling saws are designed to cut standing trees. They have a narrow, ribbonlike blade. For felling, you want a light saw with a narrow blade so that the kerf—the space created by the blade—can be wedged open early in the cut. Bucking saws, used for cutting up logs, rest on the log and have a heavier blade that gives greater cutting power without tiring the sawyer. The width of bucking saws varies from approximately 4 inches at the handles to about 6 inches in the middle.

One reason many people don't like crosscut saws is that the majority of these tools are never properly sharpened. Dull crosscut saws probably fed the Paul Bunyan myth of a superior, axe-wielding timberman as much as anything. A crosscut saw should produce sizable curled wood chips (like little noodles) with little effort other than back-and-forth pulling. When the blade starts to bind, or you get sawdust instead of "noodles," either your blade is dull or you need to wedge open the kerf.

SHARPENING A CROSSCUT SAW. Crosscut saw sharpening seems complicated at first, but it's not difficult once you understand the principles. Working with the jumbo teeth helps to begin with, because you can see them. The cutting teeth are sharpened at a compound angle like a carpenter's crosscut saw. Proper angles must be maintained and the cutting teeth must always be higher than the rakers, so the scoring of the wood by

Photo 5.2 ***"Noodles"—shavings produced by a sharp crosscut saw.***

the cutters precedes the scooping action of the rakers. All the teeth of the cutters should be one height. Setting the height is called jointing. The cutters must also be set—bent to alternating sides of the plane of the blade. To keep the blade in good condition, you'll need several specialized tools: a jointer and a raker pin gauge (sometimes combined into a single tool), a hammer set or hand anvil, a setting or swaging hammer, a spider gauge and a crosscut file. To sharpen the saw, follow these steps:

1. Make sure the gullets are deep enough. After the teeth have been filed down a number of times, the spaces between them are eventually reduced to the point that they get packed with sawdust, binding the saw. File out the gullet with the rounded back of the crosscut file or a chain saw grinding wheel. This procedure is often called "gumming."

2. Check the rakers. The edges of the rakers are supposed to flare forward, so they scoop the chips. Continual filing results in a loss of this angle. Put the saw in a vise, leaving about ¾ inch above the jaws. Strike each raker tooth with a hammer from the inside. This is called "swaging." The blow should be at an angle to the face of the V, producing as much flare as possible without damaging the point.

3. Joint the teeth, filing them to exactly the same height. Carefully position the file in the jointer tool to match the arc of the saw teeth. File until the tool just reaches the shortest tooth.

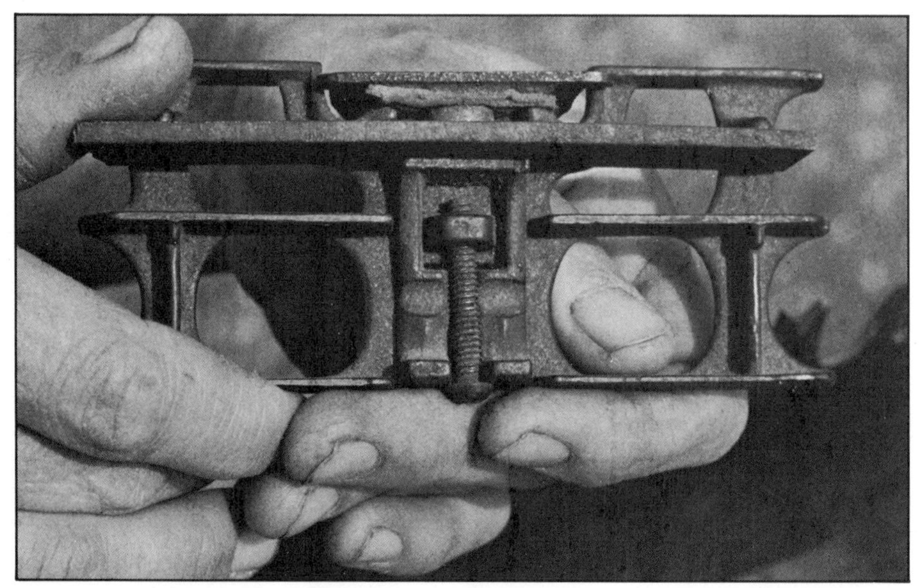

Photo 5.3 ***A typical combination jointer-raker gauge.***

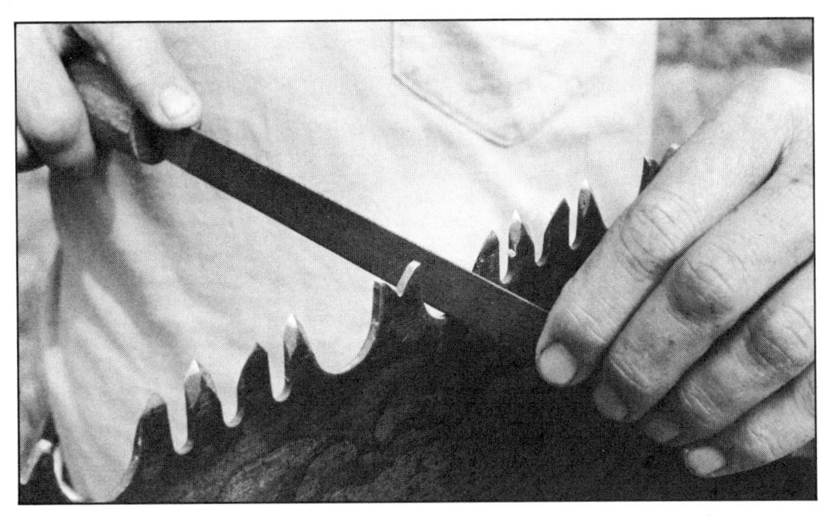

Photo 5.4 ***"Gumming"—lowering the gullets with a crosscut saw file.***

Photo 5.5 ***"Swaging" the rakers.***

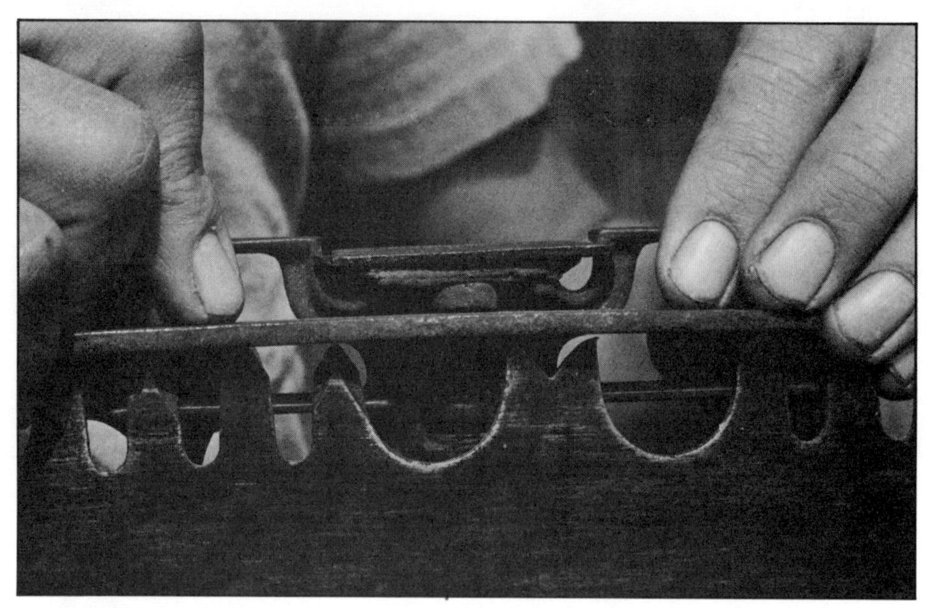

Photo 5.6 ***"Top jointing."***

Photo 5.7 ***Filing the cutters.***

4. Sharpen the cutters as with a crosscut carpentry saw, except that both edges of the teeth are filed, allowing back-and-forth cutting. File every cutter on the saw, then reverse the saw in the vise and file the remaining cutters. File away from the cutting edge, not into it. For sawing softwoods, shape the cutters into long, pointed teeth. File the angle between

the front and back slope (viewed from the side) of each tooth to 60 degrees. For hardwoods, file a shorter point, about 80 degrees. For all-around use, the angle formed by the teeth should be around 70 degrees. Stop filing when you see the flat spot caused by jointing disappear.

5. Rejoint the rakers, making sure they are a uniform distance below the sharpened cutters. Use the raker pin gauge set at a depth of 1/64 to

Photo 5.8 ***"Rejointing"—checking the raker clearance.***

Photo 5.9 ***Sharpening the rakers.***

Photo 5.10 *Using a spider gauge to check the set of the cutters.*

$1/40$ inch for hardwoods ($1/40$ to $1/32$ inch for softwoods) to make sure the distance is correct.

6. Sharpen the rakers by holding a file perpendicular to the plane of the saw blade and file the V at the middle of each tooth until the point on each side is sharp and the flat spot from jointing disappears.

7. If necessary, set the cutters by bending them outward—hammering them against a hand anvil. Set the upper $3/16$ to $1/4$ inch of each cutter to bend out $1/32$ inch, alternating the bends from one side of the blade to the other. Use a spider gauge to ensure a uniform bend. Don't set the rakers.

8. Recheck the rakers, because setting the cutters may knock them out of line.

9. Finally, rub some candle wax on the sides of the blade. You'll be surprised how much easier it is to use a nicely waxed saw.

Crosscut saws are still made in the United States and England, and there are countless old crosscuts stored away in attics and chicken coops to be had for a few dollars at most. You can usually clean up and refurbish an old crosscut easily, as long as the teeth are intact.

CHAIN SAWS. In contrast to the peaceful crosscut, chain saws are noisy and dangerous, but they do cut fast. One person can do many times the work of two people pulling a crosscut. Chain saws can also be used in confined quarters, and they can make certain cuts that are difficult with other tools. But proficiency with a chain saw requires as much practice as with any other tool.

Photo 5.11 ***Setting the cutters.***

The better chain saws sold today are dependable machines. Bar and chain maintenance must be regular, but engine repairs should be infrequent. I used my Stihl 020 for cutting fuel wood, craft supplies and building logs for over five years before it required a spark plug change or a carburetor adjustment. Fuel use is negligible, especially with the smaller saws.

When most people think about chain saws, they consider the length of the cutting bar first. Actually, bar size is secondary; the main differences between saws are in weight and engine displacement, which can range from about 1.5 to over 9 cubic inches. The mini-saws (1.5 to 2.5 cubic inches) are suitable for light jobs such as cutting firewood, limbing branches and notching logs. Bar lengths range from 12 to 16 inches. Professional loggers generally use saws that displace at least 4.5 cubic inches, with bars from 2 to 5 feet, or more. Professional saws are heavy and therefore tiring to use. In between are the middleweights, with 16- to 24-inch bars, suitable for almost any task, but cumbersome for light work and somewhat limited when it comes to serious logging.

In selecting a chain saw, first choose a brand known for uncompromising quality, and make sure you can get competent local service. Many hardware and department stores have no service facilities for the saws they sell. Delays in getting a part can result in unnecessary slowdowns.

SAFETY

Logging is hard, dangerous work. Follow standard procedures to reduce risks and don't work alone. Evaluate every logging operation carefully to assure safety, and remember that you have the final responsibility for your own personal safety.

Felling the trees for your home can be a wonderful, rewarding experience. There is special pleasure in beginning the many stages of logbuilding by selecting trees in your own woods. But logging has always been a high-risk activity and the decision to become involved in any logging activity should not be made lightly. Safety is a matter of personal attitude, procedural knowledge and proper equipment. Most accidents are caused by beginners during their first weeks on the job.

Use the buddy system, so you are always at least within yelling distance of someone else. Children, pets and casual observers should be nowhere near any logging activities. Beginners would do well to seek help from an experienced logger. Be cautious at all times.

Keep your logging tools in proper condition, comply with standard safety recommendations and use safety equipment (eye protection, hard hats and boots). Logging requires stout boots with good ankle reinforcement. The soles should provide good traction at all times. Clothing should be loose fitting for freedom of movement, but not baggy. Cut off the pants 3 to 5 inches above the ankles to avoid getting them caught. Leave the hem off, so they will tear easily in an emergency. Hard hats are mandatory. (I was hit square on my hard hat by a falling limb the first day out cutting trees for our house.) Chain saw users should also use ear protectors and face screens. Hard hats are available with ear and eye protection that pivots out of the way when not needed. Canister fire extinguishers are recommended for chain saw work in dry, brushy areas.

FELLING

The felling procedures listed here cover most felling situations, but exceptional situations should be handled with extreme caution. In questionable cases, get qualified help. Cancel all felling during windy weather.

Felling equipment, in addition to a saw, includes the safety gear, an axe and felling wedges. Felling wedges are considerably flatter and thinner than wood-splitting wedges. Chain saw wedges are made of soft aluminum or plastic to prevent damage to the chain. Homemade hardwood wedges are also useful. A small plumb bob of some sort, such as a lug nut tied to a string, is good for checking the lean of trees.

Prior to any saw work, follow these steps:

1. Check the logging site. Consider the landscape, how the trees will fall and how you will skid the logs. In any one area, all felled trees should be more or less parallel to each other. This minimizes breakage and

Photo 5.12 ***A clear escape route should be back and to one side of the felling direction.***

simplifies bucking and skidding operations.

2. Check for snags (dead trees that may fall without warning during other operations). These should be felled before other work proceeds. Because snags often have internal rotten wood, felling them is risky.

3. Clear the work area of brush, saplings and limbs. Create a clearing with room for comfortable work.

4. Determine your escape route. Have a clear place to run to once the tree starts falling, back and to one side of the felling direction. A safe escape distance for a medium tree is about 20 feet.

5. Size up the tree. Picking the lean is half the job. View the tree from several angles, at a distance from the tree, using the plumb bob as a vertical guide. Observe any predominant limbs growing on one side of the tree.

Unbalanced limbs have a considerable effect on the direction of the fall. If possible, plan to fell with the lean, but extreme leaners are dangerous and should not be felled by beginners.

6. Check the tree you plan to cut for widow makers (dead limbs that can suddenly fall), and make sure the tree won't hang up in other trees as it comes down.

7. Reexamine the escape route.

8. Walk out the lay (where the tree will fall). Check for hidden stumps, large rocks or uneven terrain that could damage the falling tree.

The steps for actually cutting down a tree are these:

1. Make a face cut by removing a wedge-shaped section from the base of the trunk. The face actually consists of two cuts: first, a horizontal cut at least one-third the diameter of the tree, then a sloping cut that terminates exactly at the end of the horizontal cut. The horizontal cut should be perpendicular to the felling direction. The sloping cut can be made from above or below the horizontal cut, but the angle should always be close to 45 degrees. The deeper the horizontal cut, the closer to vertical the sloping cut should be. A narrow face closes quickly, and the tree is in a nonguided

Photo 5.13 ***Starting the face cut. The rear handle is aimed in the felling direction.***

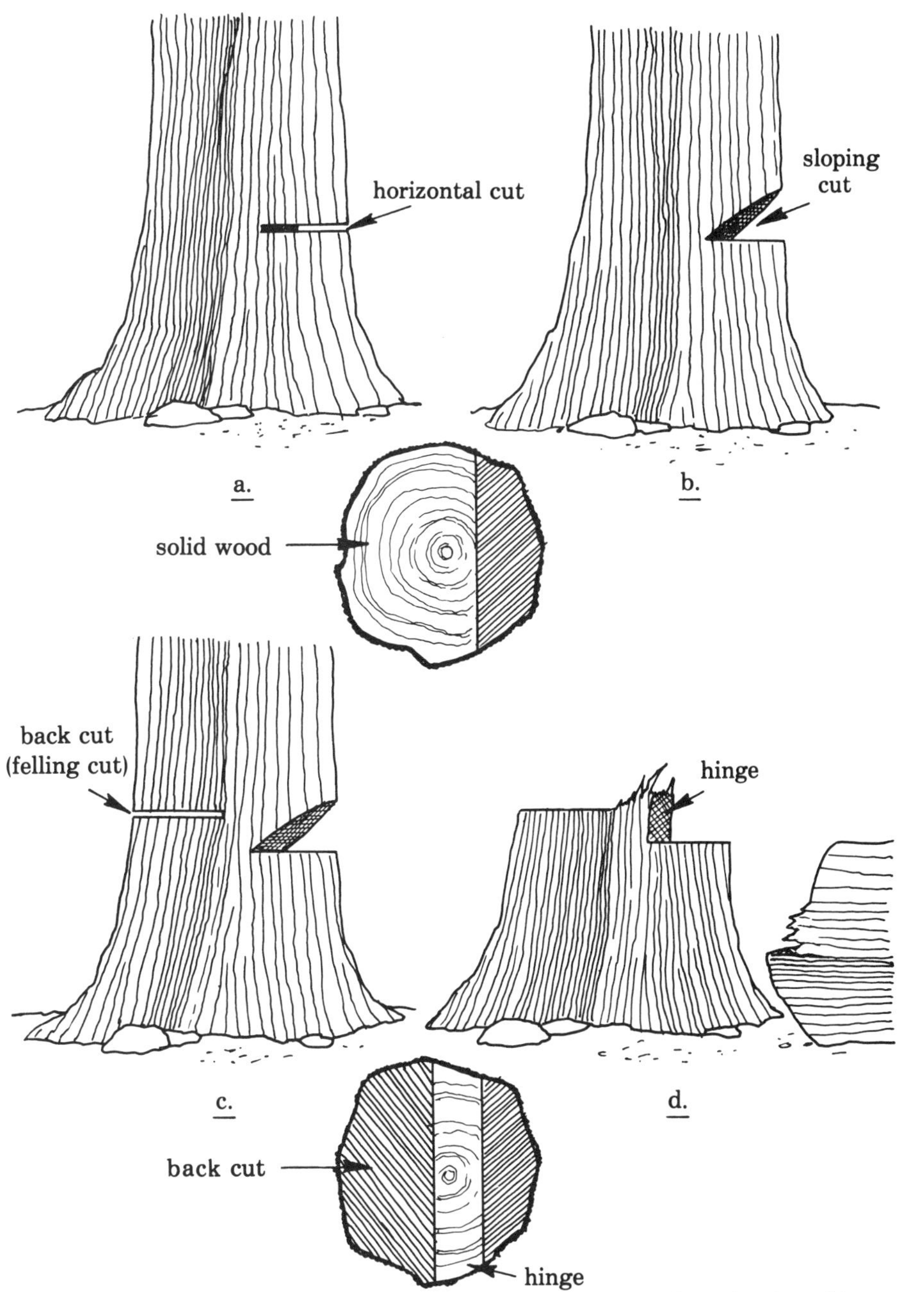

Illus. 5.2. *Cutting a tree. Make three cuts in the sequence shown above, leaving a hinge to help control the fall. Keep the back cut level, at least 2 inches above the horizontal cut.*

Photo 5.14 ***The finished face cut.***

fall as soon as the face closes. An upper sloping cut (conventional notch) saves wood because the horizontal cut is lower on the trunk. A lower sloping cut (Humbolt notch) is slightly safer because the butt drops to the ground quicker. Make your horizontal cut as low as practical, without forcing yourself into an awkward position. Trees less than 20 inches in diameter can usually be cut within 12 inches of ground level.

2. Make the back cut, or felling cut. Before starting, shut off the saw and yell a warning to make sure that no one is nearby. The felling cut should be at least 2 inches above the horizontal face cut and level. As soon as possible, insert a wedge into the kerf behind the saw. This will keep the kerf open and help to prevent the trunk from leaning back, binding the saw. Leave a hinge of unsevered wood between the face and the back cut. If the hinge is cut, you lose control of the fall. In certain cases, you can make wedge-shaped hinges to influence the felling direction.

3. Escape. Don't hesitate! Hung-up equipment should be left behind.

Trees being what they are, there will be exceptions to the ideal procedure for felling. Three of the most common problems are leaners, hangups and sit-backs.

LEANERS. If the lean isn't severe, cut the face in the desired felling direction, but alter the felling cut to leave more wood in the hinge on the side of the desired fall. The extra wood will pull the trunk toward it.

Photo 5.15 *The stump reveals an even hinge and a level back cut.*

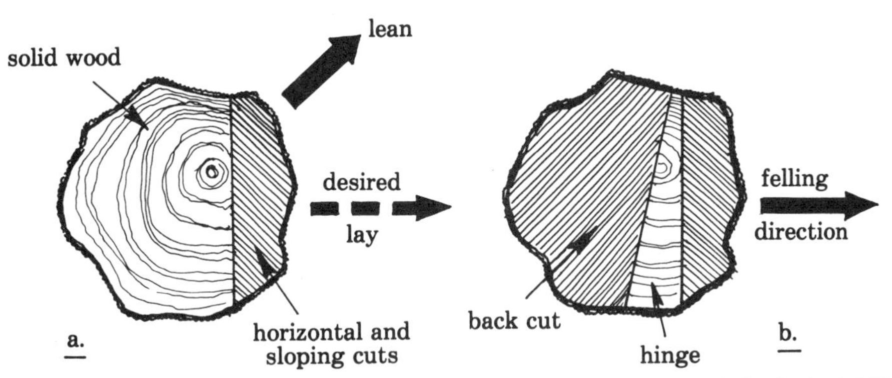

Illus. 5.3. *Cutting a leaner. (a) Cut the notch face in the direction of the desired fall. (b) With the back cut, modify the hinge area so there is more holding wood in the direction of the desired fall.*

Felling heavy leaners is tricky, professional work. Heavy leaners tend to split and do other dangerous things. Get professional help. Don't try to fell such trees.

HANGUPS. These are trees that start to fall but get stopped by obstructions. Felling the supporting tree is extremely dangerous and should never be attempted. The safest method is to fell a driver (another tree) into the hangup. If this doesn't work, *leave it*!

SIT-BACKS. These are trees that attempt to fall back toward the felling cut because the lean was misjudged. It's sometimes possible to wedge the tree forward toward the face, freeing the saw. First, check for widow makers overhead that could fall when you pound the wedges. Remove nearby bark, then use two wedges driven alternately. (Don't attempt a second cut to try to let the tree fall in the direction it is leaning; the tree could fall unexpectedly.) If wedges don't work, use another tree as a driver or get professional help.

LIMBING AND BUCKING

LIMBING, done with an axe or saw, is the removal of branches after a tree is felled. The process can be tricky and dangerous. Logs can suddenly shift when a supporting limb is cut. Clear out the area and start limbing at the butt, so the limbs slant away from you. Stay on the ground, on the uphill side of the tree, and cut flush to the trunk. Rough limbing makes logs hard to skid, scars logging roads and causes problems at the hewing site or sawmill. Always be on the lookout for limbs or saplings bent by the trunk. *Leave them*! The force released when one is cut can be tremendous. When the rest of the limbing is done, roll the tree and cut the trapped limb.

BUCKING refers to cutting the tree into logs, the decisive task in log utilization. For logbuilding, I follow a standard procedure before bucking any logs to length:

1. Make an inventory of all logs felled. Number the trees with a timber crayon and measure the usable length, the butt diameter and the diameter of the usable tip. Note special characteristics in a small notebook: straightness, bowing, twist (seen in the bark) or excessive knots.
2. Prepare a schematic drawing of all numbered trees, with all measurements included.
3. Work out the optimum utilization on paper. Start with a complete materials list of your home, filling in items as they're covered by the logs you've cut. Add at least 6 inches of cut-off waste to each log length and be sure to include two wall thicknesses plus two notch extensions (the distance the log goes past the notch). Bottom courses use the largest-diameter logs. Size progressively, reducing log diameter as the walls go up. The plate logs should be the best available, with no bow or twist. Assign a number to each log and add the abbreviation: W-wall, J-joist, S-sill, Pl-plate, G-girder,

Photo 5.16 *Limbing. Whenever possible stand uphill and on the opposite side of the log.*

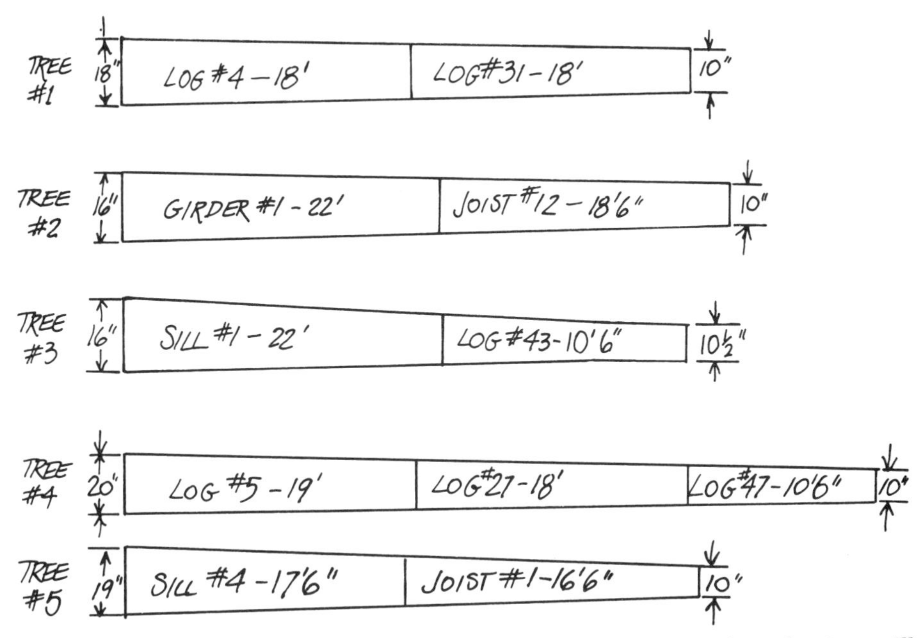

Illus. 5.4. *Make a tree-utilization chart for trees you have cut, noting where the pieces will be used in the house. This also requires a materials list, as in table 5.3.*

Table 5.2 **Sample Tree Inventory (from the field notebook)**

Tree #	Length	Butt	Tip	Comments
1	36′	18″	10″	curved, lodged
2	40′6″	16″	10″	excellent
3	32′6″	16″	10½″	good
4	47′6″	20″	10″	fair, curved
5	34′	19″	10″	good

Table 5.3 **Sample Materials List for a Log House (partial list)**

Use	Log #/Length	Tree #	
Girder	1–22′	2	
Sill	1–22′	3	
Sill	2–22′	27	
Sill	3–17′6″	19	
Sill	4–17′6″	5	
Log	1–22′	33	Course 1
Log	2–22′	13	Course 1
Log	3–15′6″	18	Course 1
Log	4–18′	1	Course 1
Log	5–19′	4	Course 2
Log	6–22′	22	Course 2
Log	7–15′6″	32	Course 2
Log	8–18′	16	Course 2
Plate	1–30′	11	
Plate	2–30′	28	
Joist	1–16′6″	5	
Joist	2–16′6″	14	
Joist	3–16′6″	20	
Joist	4–16′6″	24	
Joist	5–16′6″	11	
Joist	6–16′6″	12	
Joist	7–14′6″	6	

Table 5.4 **Sample Cutting List (from the field notebook)**

Tree #	First Cut	Second Cut	Third Cut
1–36′	L4–18′	L31–18′	–
2–40′6″	G1–22′	J12–18′6″	–
3–32′6″	S1–22′	L43–10′6″	–
4–47′6″	L5–19′	L27–18′	L47–10′6″
5–34′	S4–17′6″	J1–16′6″	–

Tag Abbreviations

L–Log	J–Joist
S–Sill	R–Rafter
Pl–Plate	Po–Post
W–Wall	SL–Saw Log
G–Girder	

R-rafter, Po-post. Butts of very large logs are best utilized as saw logs for making lumber. Mark them SL.

4. Return to the woods with a complete list of the log utilizations—the cutting list. Remeasure the full-length logs and mark out cuts. Buck the logs, and label both ends of each log with a timber crayon.

Bucking deals with two powerful forces: tension and compression, which result from uneven pressure on a log. When a log is supported in only one place, with part of it hanging in space, the underside of the hanging section is compressed by its own weight (compression), while the top portion is stretched (tension). When a log is supported in two places and unsupported in the middle, the top is under compression while the bottom is under tension. Both forces must be accounted for in making a cut. Gravity acts immediately when a log is freed from the supporting points. Logs drop fast. Trees can shift or pivot. Before cutting, check nearby trees for possible widow makers, clear out the area and establish escape routes. If possible, begin cutting from the tip toward the butt. This will often gradually relieve pressure on the logs. Always stand on the uphill side of the log. The five possible types of pressure on logs are determined by the way they are supported. Each requires a specific cutting procedure.

NO BIND occurs when the log lies flat on the ground. (1) Make a vertical cut on the side away from you, keeping the blade tip above ground level. (2) Saw straight down. The offside cut makes it possible to stand farther back during the final cut. Drive a wedge as soon as possible. Insist on a safe stance *on the ground*, and hold the chain saw to one side so you are out of the line of the blade if it kicks back.

TOP BIND results when the log is supported at two points. Straight bucking would immediately pinch the saw. (1) Make a vertical cut on the far side. (2) Take a wedge out of the top. (3) Make a vertical cut on the near side. (4) Make the final cut by bucking upward from below.

BOTTOM BIND occurs when the log to be bucked hangs unsupported in space. (1) Cut a bottom wedge to relieve compression. (2) Make a vertical cut on the far side. (3) Make a vertical cut on the near side. (4) Finish by cutting down from above.

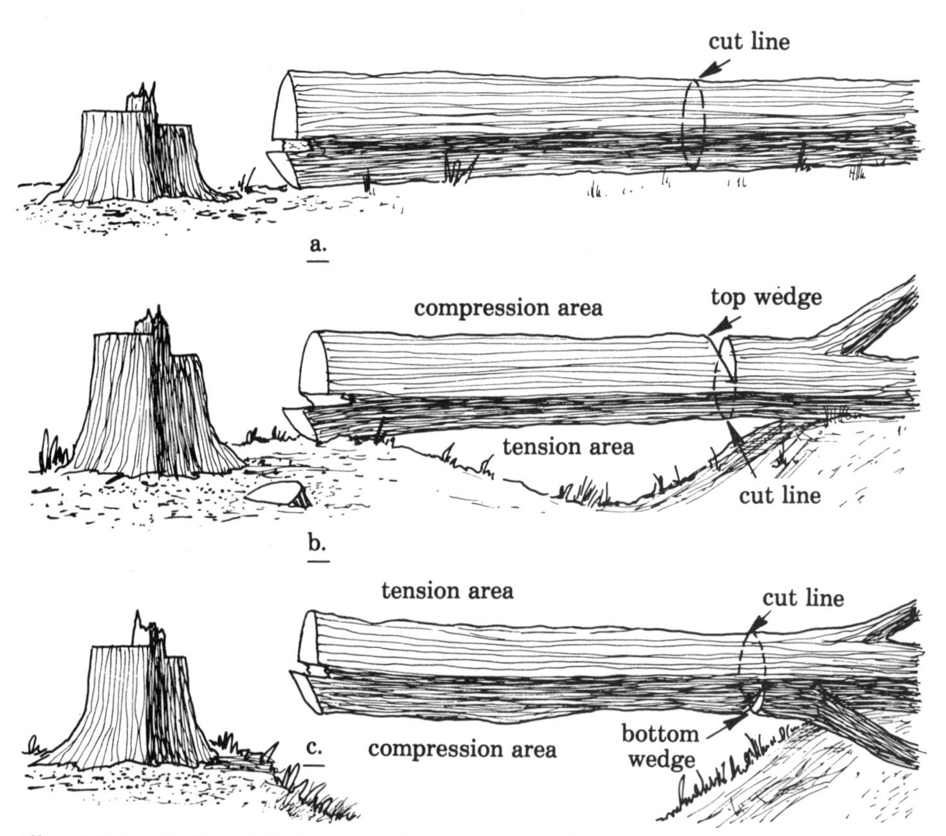

Illus. 5.5. ***(a) No bind. The log is evenly supported on the ground. Keep the chain saw off the ground. You may need to roll the log with a peavey to cut the last bit on the bottom, where it rests on the ground. (b) Top bind. Suspended log results in compression at top, tension at bottom. Cut in four steps, as indicated in text. (c) Bottom bind. Overhanging but with tension on top, compression on bottom. Cut as indicated in text.***

Photo 5.17 ***"Top bind"—undercutting a log supported at two points. Note the wedge already taken out of the top.***

END BIND is caused when a log is lying flat on a slope. To prevent it from sliding into the saw bar as the cut is completed, insert a wedge into the kerf as soon as possible.

SIDE BIND results when a felled tree is sprung among three points, as when it falls between several trees growing close together. The situation is dangerous because the log may snap back after the final cut. Get professional help.

SKIDDING

Skidding, moving the cut-up logs, may be accomplished with a tractor, a jeep with a power winch, even a team of horses. It is just as dangerous as timber felling, though. Logs often roll or shift unpredictably, and chains and cables can fail, sometimes lashing back when they break. The force of moving logs can easily crush bones or equipment. Tractors can turn over. Start out by working with an experienced logger, and always use extreme caution.

Most commercial loggers use very expensive equipment, such as skidders (mobile cranes equipped with powerful winches and long cables), high-lead systems, and skyline rigs. Ground work is done with specially rigged crawlers. In our area, some farmers do a fair amount of small-scale

Photo 5.18 ***Tractor skidding.***

skidding with standard tractors. The new four-wheel-drive tractors are well suited to occasional skidding. Tire chains added to them are also helpful for gaining traction in slippery situations, but chains can really dig up the forest floor. Perhaps the most satisfying method of skidding is to use draft horses, mules or oxen. An experienced neighbor told me that our 1,500-pound mare could outpull his 35-horsepower tractor in mud or slippery conditions on hilly terrain. But because of the skill required of the teamster, logging with draft animals should be practiced only with experienced help. The risk to loggers and animals is too high otherwise.

For small-scale skidding, you'll need logging chains 12 to 16 feet long, with 5/16- or 3/8-inch links (see photo 9.3). Choker cables are sometimes used instead of chains. For general purposes, a 3/8-inch, 12-foot cable is adequate. Log-grabs and skidding tongs save a lot of time in connecting logs to the pulling rig, and peaveys or cant hooks are indispensable for rolling, lifting and prying logs. A tractor with a lift boom (a simple crane of welded pipe) is also very useful. You can accomplish short hauls with the aid of a timber

Photo 5.19 ***Rolling a log with peaveys.***

cart, a device consisting of a set of wheels with an axle between, where the log rests (see photo 9.2), or a timber carrier, which consists of grab hooks connected with a swivel to a 4-foot wooden bar and hooks at both ends.

Skidding is tricky. Green logs are incredibly heavy, and uphill pulling is difficult. On the downhill slope the log is hard to stop. Skidding along the side of a slope can be dangerous also, because the logs may roll downhill, twisting the chain and pulling the tractor or animals out of control. The first requirement for skidding is to connect the chain to the log. Wrap the end of the chain that has a slip hook attached around the big end of the log, then around itself and back to where it loops under the log, and secure it with the slip hook. Attach the other end of the chain to the tractor. With tongs or grabs you can sometimes eliminate the chain, although it's important to have one available. Log tongs are nice to use because you don't have to get under the log, and they're self-tightening as the pulling force is applied.

A good skidding road follows the most level path out of the woods. For downhill routes, the best option is often a straight descent. Such

Photo 5.20 ***Lifting a log with two crossed peaveys.***

pulls on snow or slick mud, however, are dangerous. Sometimes two small logs can be chained side by side so that they act as brakes against each other, preventing the logs from rolling sideways. To chain the logs in this manner, wrap and secure the opposite end of a single chain around each log. Hook a second chain between the logs, and pull from the middle. For easy pulls, another method is to connect several logs end to end and snake them out in one run, single file.

The 100-foot cable and pulley is another worthwhile device, a combination that's useful where a tractor can't go, up steep slopes, across gullies or over very wet areas. The pulley keeps the cable off the ground, and makes it possible to pull at an angle (or downhill) to the direction that the log is being moved. You can attach the pulley to a tree where a path bends and actually pull around the corner. Back the tractor up as far as you can toward the bend, then pass the cable through the pulley and secure the end to the log. You can now pull the log toward the tree where the pulley is fastened, by driving the tractor away from the bend.

Photo 5.21 *A self-tightening chain hitch.*

Photo 5.22 *Using skidding tongs linked to the drawbar on a tractor hitch.*

Illus. 5.6. *Using a cable with a pulley to skid a log.*

LOG STORAGE

Storage should be organized both to prevent log deterioration and to permit access to the logs. The storage area should be near the place where logs will be hewn and notched. A good work area is level and preferably under a fine shade tree. Aim to minimize moving logs by hand. Arrange the logs so they can be rolled easily into place as needed. In Switzerland, small sawmills often use a winch and cable to pull logs out of storage piles. Then an overhead track and trolley lifts the logs onto the saw carriage. A similar system would be ideal for log hewing.

Bark and wood chip storage arrangements are also necessary and should be worked out in advance. Large bins several inches off the ground are excellent. Plastic tarps can be used as covers. In our case, the bark piles got out of control; huge mountains built up and the bark subsequently

Photo 5.23 ***Numbered logs in storage.***

became so waterlogged that it wouldn't burn. A storage area for finished logs is usually necessary, too.

The basic idea is to keep logs well off the ground with air space between them. Use sound, seasoned cross logs as spacers. Bowed logs should be stored crown up. An open shed or drafty, unused barn is excellent. Some logbuilders recommend storing coniferous logs with the bark left in place to protect against fungi—potential insect damage being considered preferable to fungi invasion. But deciduous trees should be barked by early spring, then kept shaded. Do not store logs under plastic sheeting, which holds moisture. On sunny days, the logs will heat up, offering prime conditions for fungi, which thrive in a humid environment between 75° and 90°F.

CHAPTER VI

FOUNDATIONS

When you're eager to start hewing logs, it's easy to forget the importance of a good foundation. Builders often consider foundation work to be unrewarding, difficult and time-consuming. But the foundation is the base of a building. The permanence of a log structure depends on a good foundation and protective roofing.

Most foundations consist of concrete footings supporting a continuous stem wall, but you can also make a foundation of concrete or stone piers spaced under the log wall. You'll still need footings if you use piers. The footings are broader than the stem wall or piers and distribute the weight of the building over a sufficiently wide area to prevent settling or tilting. The stem wall rests on the footings and usually extends a foot or so above grade (ground level) to keep the log walls from touching the ground. Because some kinds of soil support weight better than others, the size of the footings is determined by soil type as well as by the size of the house.

CONCRETE AND MORTAR

Concrete is a mixture of sand, graded gravel, cement and water that is used for poured foundation footings, stem walls and slab floors. Mortar is the same thing without the gravel. Mortar is used for joining masonry blocks and stones in masonry walls. What concrete and mortar have in common is cement, a special mix of limestone and other materials. This mixture is heated and pulverized into a powder that, when mixed with water, binds tightly with sand and gravel to form a hard, permanent substance that has proved invaluable in construction work.

Poured concrete is very strong in compression—which is the basic force exerted by house load. Under tension, however, it breaks quite easily, and many states therefore require steel reinforcing bars (called rebar) to be embedded in concrete foundations for extra strength. For the small extra cost, rebar is well worth installing.

Concrete and mortar cure by the chemical process of hydration, not by mere drying. The water used in the mix bonds chemically with the other materials. It's important to use only as much water as required to make a paste that will coat the aggregate (sand and gravel) in the mixture, because excess water leaves minute channels that reduce strength. For poured concrete, 6 gallons of water per cubic foot of cement is about right.

Photo 6.1 *A poured concrete foundation. Similar excavation, batter boards, and footings are recommended for masonry foundations.*

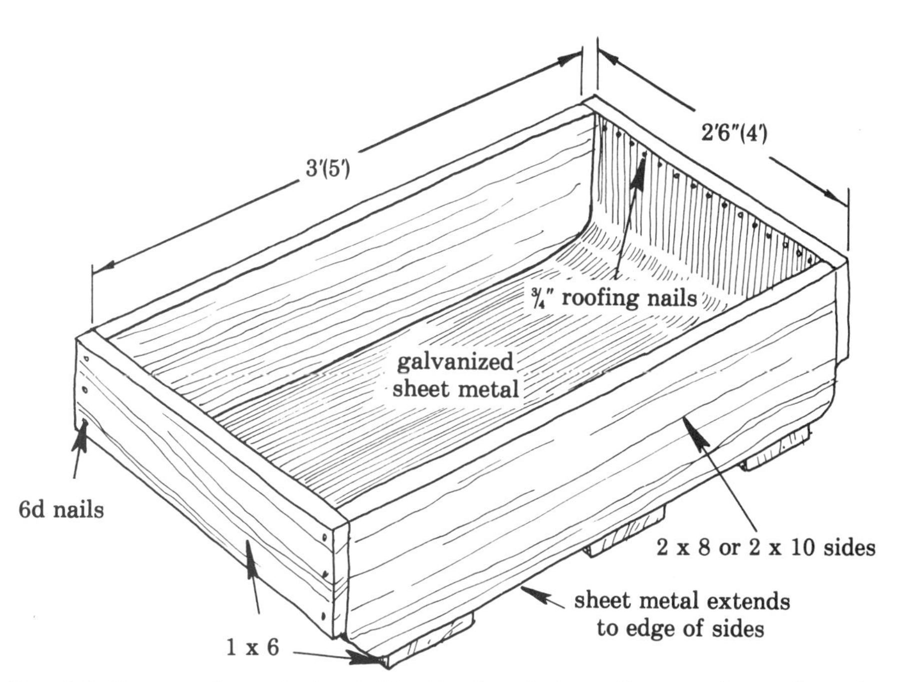

Illus. 6.1. *Homemade mortar boat. Specifications in parentheses are larger dimensions for mixing concrete. Bottom is galvanized sheet metal.*

The cement, to mix well, must be a free-flowing powder with no lumps, and the aggregate should be clean and hard. Rocks up to 6 inches across can also be embedded in a concrete pour. The dry portion must be thoroughly mixed before the water is added. The ratio for concrete is usually 1 part cement to 2 parts sand and 3 parts gravel. For masonry it is 1 part lime to 2 parts cement and 9 parts sand. The lime helps make the mortar more workable. Use Type I Portland cement for concrete and Type B (if available) for mortar. Before you make a pour, wet the forms and the earth at the base of the trench to prevent moisture from being drawn away from the concrete.

For hand mixing, at least four complete turnings are usually required. A special cement-mixing hoe, with two holes in the blade, is helpful. For concrete pours, hand mixing is slow, hard-on-the-back labor. When we hand mixed concrete for our footing and below-grade stem wall, three of us worked 1½ days on the job. Small cement mixers can mix a batch of concrete in a few minutes. Gas-engine mixers can be rented, or a generator can be used to power an electric mixer. Cement mixers mounted to the three-point hitch of a tractor can run off the power takeoff. These units are very useful at remote construction sites. With mechanical mixers, careful daily cleaning is mandatory. Wash the mixer with water, then use coarse aggregate to scour the drum. Do not pound metal parts to loosen hardened concrete. Concrete sticks to dents. All tools should be cleaned after each work session.

Transit mix (sometimes called ready mix) may be as economical as home mixing. It is delivered in trucks carrying up to 5 cubic yards of concrete. There is usually a minimum order, plus charges for mileage and time at the site. Concrete trucks are very heavy, so you must provide a good road to the site, and an adequate turn-around area.

Concrete begins to set in about an hour. After three days, forms should be loosened, but left in place. Wet burlap can be draped over exposed areas. Final set takes seven days, and final curing takes up to a month to achieve full strength. Do concrete and masonry work in mild weather. Both freezing and hot weather interefere with hydration. On a sunny day, you can prevent excessive loss of water by covering the concrete with wet burlap sacks.

PREPARING THE SITE

Log houses look best in a natural setting. This usually means that grading should be kept to a minimum. You can dig a trench for the foundation without changing the slope at all, or level only the building site, but the best drainage results from having the land slope away from the house slightly in all directions. If this is not possible without major grading, install a drain system on the uphill side.

Clear away debris and underbrush from the immediate building area, but protect trees that are to be saved with burlap sacks or wrap-around fences. Don't excavate or fill at the base of living trees. The foundation itself should not rest on any topsoil, loose fill or organic material.

Pour the footings below the local frost line to prevent frost heaving (when the ground surface freezes in winter, it expands, making any foundation resting at this level unstable). In our area, 2 feet deep is below the frost line. Check with local builders because some small pockets are much colder than the surrounding area. The base of the footings should be level, requiring stepped construction on a sloped site. Footings generally range from 8 to 10 inches thick and 16 to 24 inches wide for medium-sized buildings on most types of soil. The smaller dimensions should be adequate for a single-story log house built on compact, well-drained subsoil. Two or three horizontal reinforcing bars embedded in the concrete can be used for additional strength. Foundations constructed on potentially wet sites, such as basins or watersheds, must be protected with perforated drains placed around the footings. Black 4-inch-diameter corrugated polyethylene pipe has virtually replaced conventional ceramic pipe for such drains. This plastic material is lightweight, inexpensive and very resistant to damage or decomposition. Drainpipes should be covered with gravel fill.

SIZING THE FOOTINGS

You can roughly calculate a limit for your footing dimensions. It's generally advisable to construct a foundation capable of supporting twice the calculated building load. In questionable circumstances consult local codes and home contractors in the area. Subsoils for foundations can be clay, sand, gravel or rock. Each type has a different bearing capacity, as indicated in table 6.1.

Table 6.1 **Bearing Capacity of Different Soil Types**

Soil Type	Total Bearing Capacity (in pounds per square foot)
Soft clay	2,000
Fine sand, firm clay	4,000
Loose, coarse sand	6,000
Compacted, coarse sand	8,000
Soft rock	16,000
Hard, sound rock	80,000

Building load is divided into dead and live load. Dead load is the weight of the building—including the heaviest element, the foundation. Live load includes the occupants and their possessions, plus snow and wind factors. You've got to account for both to be sure the subsoil can support the total load. Loads vary considerably for different construction methods and materials, but you can make a reasonable estimate by consulting architects' tables for various building materials. For our house, I estimated loads at the levels indicated in tables 6.2 and 6.3.

Table 6.2 **Estimates for Load-Bearing Materials (for Langsner house)**

Item	Load Exerted
Concrete footing (8 × 24 in., 125 lb./cu.ft.)	160 lbs./linear ft.
Concrete and stone stem wall (12 × 48 in. average)	500 lbs./linear ft.
Log walls (6 in. thick)	20 lbs./sq. ft.
Flooring (2 in. thick)	6.6 lbs./sq. ft.
Joists (6 × 9 in.)	13 lbs./linear ft.
Roof (including rafters)	15 lbs./sq. ft.
Occupants, contents	50 lbs./sq. ft.
Snow, wind load (hypothetical maximum)	50 lbs./sq. ft.

Table 6.3 **Sample Building Load Calculation (for Langsner house)**

Item		Load Weight
Footing (160 lb. × 72 linear feet)	=	11,520 lb.
Stem wall (500 lb. × 72 linear feet)	=	36,000 lb.
Walls (20 lb. × 1,152 square feet)	=	23,040 lb.
Flooring (6.6 lb. × 960 square feet)	=	6,336 lb.
Joists (13 lb. × 336 linear feet)	=	4,368 lb.
Roof (15 lb. × 840 feet)	=	12,600 lb.
Occupants, contents (50 lb. × 640 feet)	=	32,000 lb.
Snow, wind (50 lb. × 840 feet)	=	42,000 lb.
Total Load		167,864 lb.

To see if our house would be supported adequately by the clay soil in our area, I took the total estimated weight of the building and its contents (from table 6.3, 167,864 pounds) and divided by the footing area of 144 square feet for a figure of 1,166 pounds per square foot. I doubled this to 2,332 as a safety factor and compared that with the bearing capacity of our soil (4,000 pounds per square foot, from table 6.1). Thus, the proposed house was well within acceptable limits.

LAYING OUT THE FOUNDATION

Layout is critical to the entire building process. Corners must be square and properly spaced from each other. You don't need survey tools to do a foundation, just a tape measure as long as the diagonal of the structure, a small plumb bob, and an image level or a water level. A water level can be made from clear plastic tubing mostly filled with water, with the ends plugged. Do not attempt to use a hanging line level because its

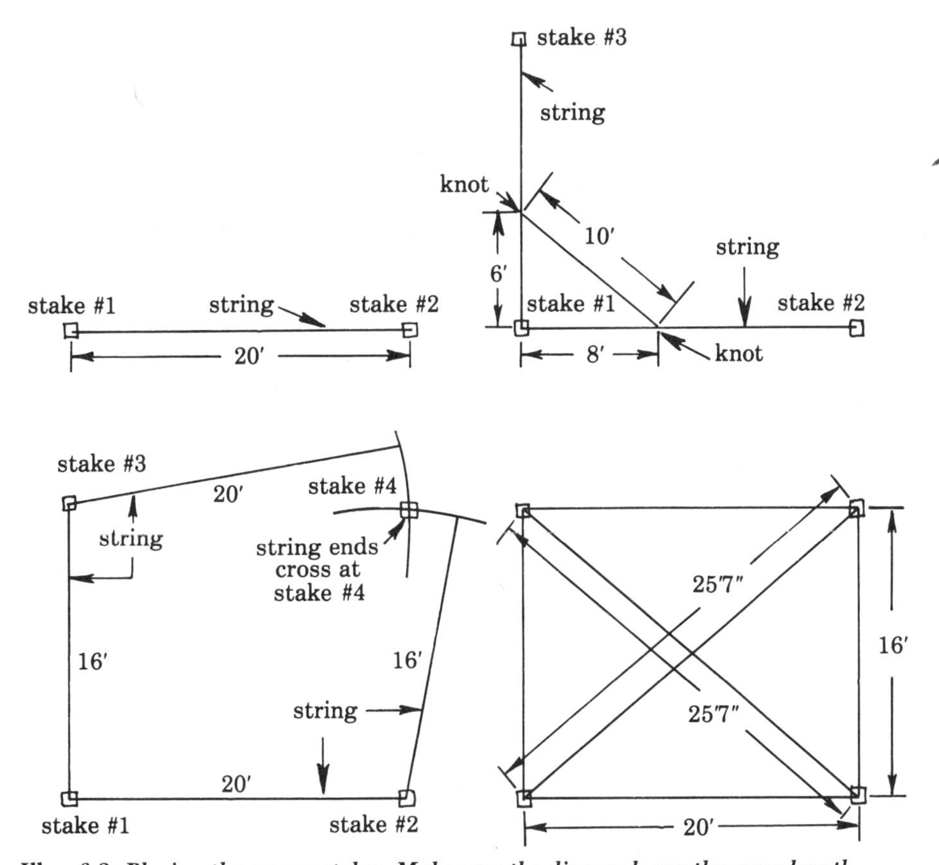

Illus. 6.2. *Placing the corner stakes. Make sure the diagonals are the same length.*

weight would cause the string to sag, throwing off the measurement.

Outline the foundation area roughly. Lime is useful as a marker—that's what they use to mark football fields. Remove all topsoil and set it aside for use in replanting groundcover after construction is finished. Try to relocate any sod. The stripped foundation area will look rough, but there is no need for organic matter or topsoil under a house.

Lay out the corners for the foundation by driving stakes at each corner. Drive a nail into each stake where the outermost corners of the foundation wall will be. If you want to orient the foundation (and thus the walls of the completed house) along a particular compass direction, use a magnetic compass and sight along the angles marked on its face to set down the first foundation line. The actual location of the second stake should be measured with a tape rule. This measurement should equal the planned distance between the outer corners of the foundation wall. Tie builder's twine between the nails in the first two stakes. Locate the third stake so that a string stretched between the first and third stakes is at right angles to the string fastened between stakes one and two. Use a framing square to line up the string. Drive the third stake temporarily, the proper distance between the second set of corners. Then fine-tune the angle by taking advantage of the Pythagorean theorem: the hypotenuse of a right triangle equals the square root of the sum of the squares of the other two sides (square a number by multiplying it by itself). Measure out 6 feet along one string and 8 feet along the other and tie knots at each point. The sum of the squares of these distances is 36 plus 64, or 100, of which the square root is 10. Simply adjust the direction of the second string slightly, until the distance between the two knots is exactly 10 feet. Locate the fourth stake by swinging arcs with string from stakes two and three. If the strings are knotted at the intended distance between foundation corners, the arcs will intersect at the proper place. Check the result by measuring between the opposite corners of the foundation. If they are equal within ¼ inch, the foundation is considered square.

To mark the location of the corners without having the stakes buried by the foundation, build batter boards several feet back from each corner by driving into the ground three 2 × 4 stakes arranged in a rough right triangle. Between these, nail two horizontal ledger boards per corner to hold the string. If necessary, brace the stakes to keep them steady. You'll need 12 stakes and eight ledger boards for a rectangular foundation, more for additions to the basic rectangle. The top edges of the horizontal ledgers should be level and all at the same height, slightly higher than the top of the intended foundation wall. Notch the ledger board under each string to secure the layout lines.

EXCAVATION

You can dig out the foundation by hand or hire a backhoe. For a full basement, you'll obviously need the machinery, but even a modest trench for the footings can be laborious by hand, especially in rocky soil. Our 72-foot perimeter took four days to dig. A backhoe could probably do the same work in about an hour. However, you may do a neater job by hand, which is important if the footings will be poured without forms. It may also be difficult to get a backhoe to a remote site. Whatever method you use, try to get all excavation out of the way at once, including any trenches required for plumbing or electrical systems.

If you are building wooden forms to hold poured concrete for a footing, dig the trench wide enough to work in—excavate an additional foot on each side of the form. For a full basement, allow 2 feet of work space on the exterior sides of the forms. Keep the dirt in neat piles for use as fill later on. Trenches deeper than 4 feet must be shored if there is any chance

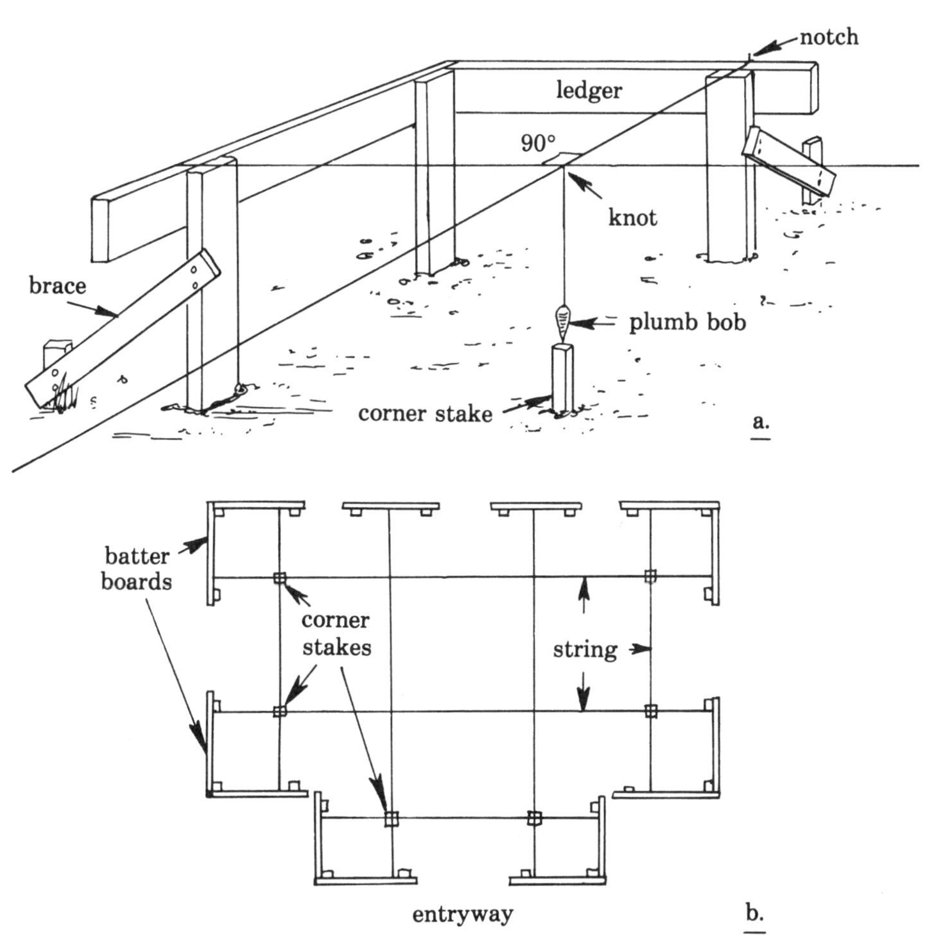

Illus. 6.3. *(a) One set of batter boards. (b) Full layout scheme for a house with an entryway.*

of cave-in; a worker buried only to the chest can quickly suffocate. The bottom of the footing must be flat to distribute weight evenly. Don't be surprised if the area enclosed by the foundation trench looks very small. Your house will seem to grow as you build it.

FOOTINGS AND STEM WALLS

You can either pour footings directly into a cleanly dug trench or into wooden forms set in the trench. A direct pour will require more concrete, but the time saved by not having to make the forms may make that a better choice—if the soil is of the right consistency to be dug cleanly.

For separate pourings of footings and stem wall, let the footing set for 48 hours, then add the stem wall on top. You can also pour both at the same time. A poured stem wall must have wooden forms, because it extends above grade. You can also build a masonry stem wall (rocks and mortar) or one of concrete blocks. A masonry wall looks just right with logs, but making one requires skill and time. If you are willing to tackle a masonry stem wall, you might consider running it up all the way to the bottom sills of the windows, a building style that emphasizes the masonry base even more.

Stem wall design may include features relevant to log construction. Often two opposing stem walls are terminated at half the sill log diameter below the adjacent pairs of opposite walls. This adjusts for height difference caused by notching, and permits round sill logs to rest on the stem wall. Some builders cast a ledge or pockets in the stem wall to support the floor joists.

Illus. 6.4. ***Footing form.***

FOOTINGS

If the trench cannot be excavated neatly, or if you plan to install a drainage system around the perimeter, build forms of ¾-inch plywood or 1-inch-thick lumber, braced outside with 2 × 4 stakes driven into the ground. To make it easier to remove the forms when the concrete has set, use duplex (double-headed) nails, or drive the nails in only partway. Nail from the stakes into the form boards, so the nail head won't be buried by concrete.

For extra strength, the footings should have several steel reinforcing bars laid horizontally several inches above the trench bottom. When the footings and stem walls are poured separately, embed steel reinforcing bars vertically in the footing as well so that they extend above it, to help secure the stem wall when it is added. Some codes also require a channel (or keyway) to be cast into the surface of the footing. This is done by placing 2 × 4s with tapered sides into the top of the footing before the concrete sets. Afterward they are removed, leaving a trough. The sides of each 2 × 4 should converge toward the base of the footing.

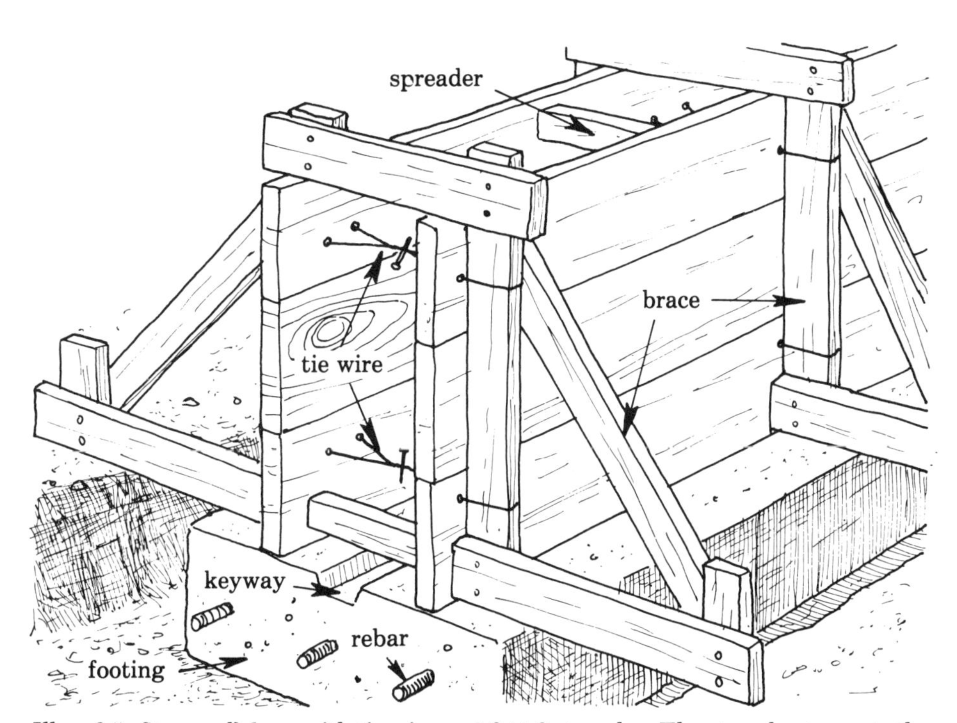

Illus. 6.5. *Stem wall form with tie wire and 2 × 2 spreader. The spreader prevents the walls from collapsing during the pour; the tie wire keeps the walls from being pushed apart by the concrete.*

POURED STEM WALLS

Poured stem walls are generally twice the thickness of the building walls. A 12-inch stem wall is adequate for a 6-inch or wider log wall. Wall forms must be strong and well braced to resist the heavy side pressure of wet concrete. Forms up to 4 feet high can be built with 1-inch-thick boards, supported by 2 × 4 stud braces spaced 2 feet apart. These are often joined to additional stakes set back from the wall for extra support. For higher walls, use 2-inch sheathing or exterior plywood, and nail horizontal 2 × 4s, called wales, to the outside of the braces to prevent buckling during the pour. Forms built to the sill line should be carefully leveled.

Space tie wires and 2 × 2 wood spreaders (or commercial snap ties) every 2 to 3 feet along the forms. Plastic pipe can be set between the forms to house plumbing and underground wires which will pass through the foundation wall. Brushing the forms with a light coat of oil before the pour (used motor oil is most readily available) makes their removal somewhat easier, but is not a crucial step. Knock out the wood spreaders during the pour, but leave the tie wires. As you pour, embed horizontal reinforcing rods at each 12-inch interval of height. To prevent air spaces, called honeycomb, tamp the concrete or vibrate the woodwork frequently during the pour.

MASONRY FOUNDATIONS

Masonry foundations are made with stones or concrete blocks, usually on a regular poured concrete footing. You can build a stone foundation without mortar—carefully laid stone walls on a solid base are strong and permanent—but dry stonemasonry such as this requires a great deal of practice to master. Mortar more easily fills gaps between stones than the rock shims used by dry-masons. Though mortar is not glue—it does not bind stones together—it does provide for the solid, stable fit of one stone upon another, even though it is often possible to pick the top stone off the bottom one even after the mortar has set. Clay has traditionally been used as mortar. It can work very well, but eventually it washes away. Mortar made by mixing lime and sand is also traditional. This type also withstands weathering better than clay. Portland cement makes the best mortar, and is recommended since it is not only hard and weather resistant, but also adds somewhat to the foundation's structural strength.

Stonemasonry is hard work, and the art must be developed with practice. Careful selection of stones from the fields pays off in time saved during construction. Stones with flat planes and square corners are much easier to lay than those with irregular shapes. Hauling rock requires time and effort and is hard on vehicles. Stones should not be too heavy. For practicality, their weight should be limited to whatever can be easily handled by two workers. Each stone must be clean—scrubbed, if necessary—

and laid on its broadest face. Place the larger stones on the lower courses. Stones should lie flat, not sloped. When you lay the stones, stagger the vertical joints, since aligning vertical joints creates planes of weakness. Keep in mind that each rock you lay will form part of the bed for the succeeding course. Use wide stones periodically to tie the front and back of the wall together. Pre-wet porous stones before mortaring them in place. Stonemasons usually hand mix their mortar in small, quickly used batches. Mortar for stonemasonry should resemble damp sand.

We used back-form masonry for our stone foundation. With this method, the front face of the wall is laid much as it would be with a standard masonry wall, but the inner side of the wall is of poured concrete. This

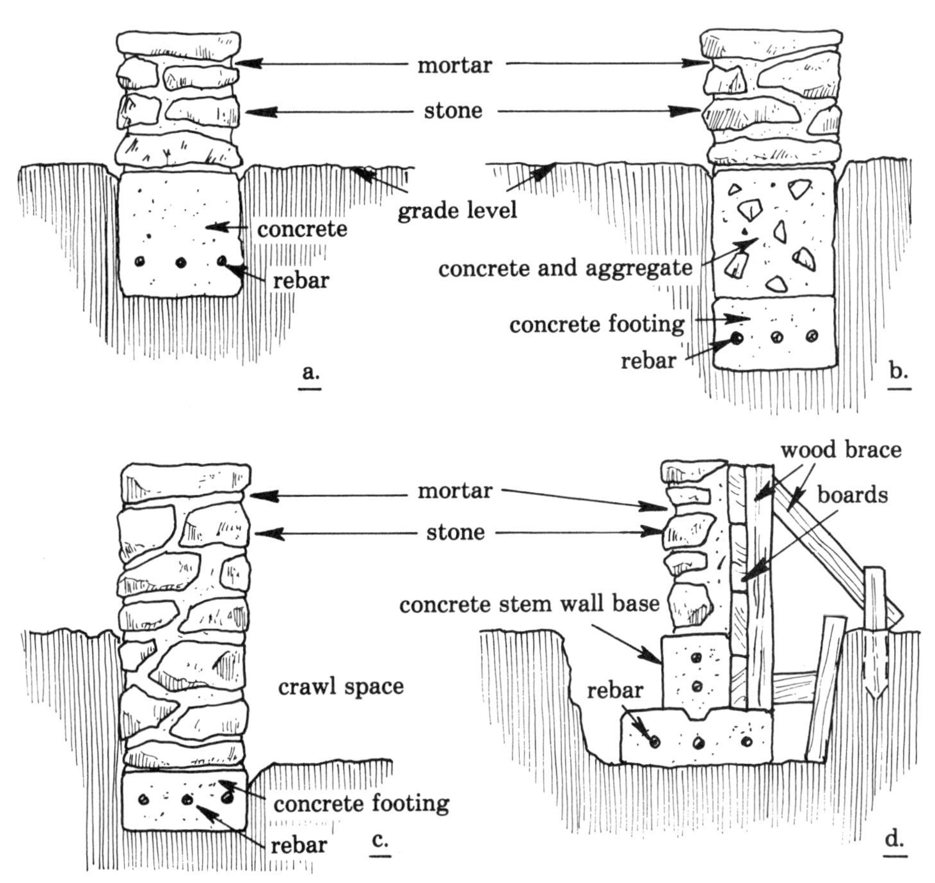

Illus. 6.6. ***(a) Concrete footing pad poured to grade. (b) Twelve-inch concrete footing. Pea concrete and rubble poured to grade. (c) Twelve-inch concrete footing with excavated crawl space area. (d) Back-forming.***

Photo 6.2 ***Back-formed masonry, stepped to accommodate crossed sill logs; the Gott homestead.***

Photo 6.3 ***Interior of back-formed masonry presents a smooth concrete surface; the Gott homestead.***

is a hybrid technique—part concrete work. It is an excellent method for amateur masons. The work is simplified because you need to pay attention to only one face. The concrete fill provides a great deal of structural strength. The method can be used on a much narrower base than would otherwise be possible with conventional masonry. Our 12-inch back-formed foundation is built on a 12-inch poured concrete stem wall.

Dress (trim and smooth) the joints for all kinds of masonry just after the mortar becomes granular. We did a fair amount of midnight scraping, by lamplight, as the mortar would have gotten too hard by morning. We used an old file as a scraping tool; it worked beautifully. Neat dressing makes a big difference in the effect of the whole job.

Photo 6.4 ***Mortarless concrete block construction. Used sheet metal roofing panels protect corner notches during nonworking periods.***

CONCRETE BLOCKS

Concrete blocks can be terminated at grade or extended to the log sills. Cinder blocks are almost as strong as standard blocks; they're somewhat less expensive and much lighter. Blocks are manufactured in nominal 4-, 6-, 8-, 10- and 12-inch widths. The nominal size of standard blocks is 16 inches long and 4 or 8 inches high. The actual dimensions of a concrete block are ⅜ inch less than the nominal specifications; the difference is generally filled by mortar.

Mortarless block construction uses a cement-and-fiberglass surface-bonding material. Two brands are Surewall and Blockbond. The first course of blocks is carefully laid in mortar, with the blocks butted together side by side, and further courses are laid dry. When the wall is complete, the surface-bonding cement is troweled onto both sides of the blocks. The end result looks like stucco, and the material is easy to mix and apply.

COMPLETING THE FOUNDATION

The foundation will be ready for the logs as soon as it is capped off and the insulation is in place. The final step is to waterproof the foundation wall below grade and fill in the excavated soil up to grade level. Your building site might not look like much at this point, but the walls tend to go up quickly, and once they do, you'll be calling it home.

THE FOUNDATION CAP

The top surface of the foundation can be capped with troweled concrete, mortar, flat rocks, or block. The cap must be level. Tie bolts aren't needed because log structures maintain their stability by their own weight and the notches in the logs. Concrete blocks can be capped with mortar by inserting wire mesh between the last two courses. Solid capping blocks are also available. Generally, no practical purpose is served by attempting to embed the sill logs into wet mortar capping. Logs will shrink and reopen the space.

INSULATION

Because masonry is a poor insulator, many energy-conscious builders are now using insulation between the outside of the stem wall and earth backfill. The warmer crawl space significantly contributes to energy conservation. Styrofoam board 1 to 3 inches thick is an excellent choice, but it must be stuccoed with mortar, and properly sealed against moisture.

Placing insulation outside the foundation allows heat to be stored in the masonry, moderating the temperature in the house, because masonry holds a great amount of heat, even though it is a poor insulator. Exterior insulation panels above grade should be covered with stucco mortar or surface-bonding cement. The cavities of cement blocks can be also filled with loose vermiculite, but not as a substitute for foam panels.

Stone walls can be insulated by building them in double courses, with a foam panel sandwiched in between. Styrofoam is recommended. The drawback is that the masonry is weakened from the lack of cross-wall bonding stones.

VENTILATION AND MOISTURE

Crawl spaces and basements should be well ventilated so that ground moisture is not absorbed by wooden girders, joists and sill logs. Sill logs in contact with concrete foundation walls should be hewn from decay-resistant heartwood, or protected with flashing or a chemical fungicide. Damp, warm woodwork is an ideal environment for fungal dry rot. At least two vents should be incorporated into each side of the foundation, preferably just beneath the sill line. Vents can be screened and louvered. Pop-in plugs (fitted insulating boards) can be used during cold weather to keep heat from escaping through the vent. Permanent vent frames (built with rough lumber) can be tacked to the concrete forms, incorporating the openings in the wall during construction. Concrete block walls often use manufactured vents that are identically sized to the blocks.

Damage from crawl-space moisture can also be prevented by covering the soil surface with polyethylene sheeting (covered with fine gravel) and treating log undersides with a fungicidal water-repellent preservative.

In damp or poorly drained soil, foundation footings should be surrounded by drain tile covered with 8 to 12 inches of clean gravel. Poured concrete stem walls can be sealed with two coats of asphalt waterproofing compound. Below-grade concrete blocks can be covered with two coats of cement plaster mortar, followed by two coats of the asphalt compound.

CHAPTER VII

HEWING

For log houses, hewing sometimes means leveling off two sides of the logs so they will lie flat as they are stacked up for the walls, but still appear round once they are in place. More often, however, it is the flat sides which are left exposed; the unhewn sides form long hollows the length of the wall into which chinking is packed. Sometimes all four sides are squared off, as in a style of building known as *pièce-en-pièce* construction, in which slotted vertical timbers form a frame and squared-off timbers are placed between the verticals to fill in the walls. Just to complete the picture there is also a style of hewing that involves flattening only three sides of the log. This style allows for a round-log appearance on the outside of the building, a flat-wall appearance inside, and flat log faces above and below for precise stacking.

A finished hewn timber will have faint marks left from the axe scoring and a slight surface irregularity, but a good hewer can cut to remarkably close tolerances. It's a skill that combines fairly strenuous chopping with the delicacy required to adhere to a chalk line. To do it well, you've got to develop a steady rhythm, but once you have that, you can go right on hewing for some time—which is what's needed when you build a house of hewn logs. Hewing is perfect cool- and cold-weather work. In summer, the work calls for a good shade tree and a nice breeze.

Whenever possible, hew, notch and raise the logs while they are still green. Seasoned logs are much harder to work, though considerably lighter. Green logs settle into place nicely, making tight joints. Some woods, like poplar, may twist or bow during seasoning, even after the logs are in final position, but, even so, the ease of working them green outweighs the inconvenience of having some logs twist.

SAFETY

To minimize the possibility of accidents, keep safety in mind while you are working and as you set up. Keep your tools as sharp as possible. Always work with a sure footing. Clear away trash and debris that could cause you to slip or lose your balance. Logs are heavy and unwieldy.

Learn and apply the techniques for moving logs. Whenever possible, move the log to a comfortable work position. If possible, store logs uphill from the hewing area, which in turn can be above the work site. Use

Photo 7.1 *Hewing. The craftsman's view.*

mechanical aids such as inclined planes, levers, wheels, rollers and block and tackle. Learn to lift with your legs, by bending your knees, rather than with your back. Debilitating back problems may sneak up on you, knocking you out of action for weeks. If you cannot reasonably do something alone, get help.

BROADAXES

Broadaxes have a bevel on one face of the blade only, similar to an ordinary carpenter's chisel. The other face is flat from the eye (where the handle is inserted) right down to the cutting edge. The chisel edge allows the axe to cut at a very narrow angle to the log, making the broadaxe perfect for hewing the log along its length. On many broadaxes, the flat side is actually slightly bowed from corner to corner of the cutting edge.

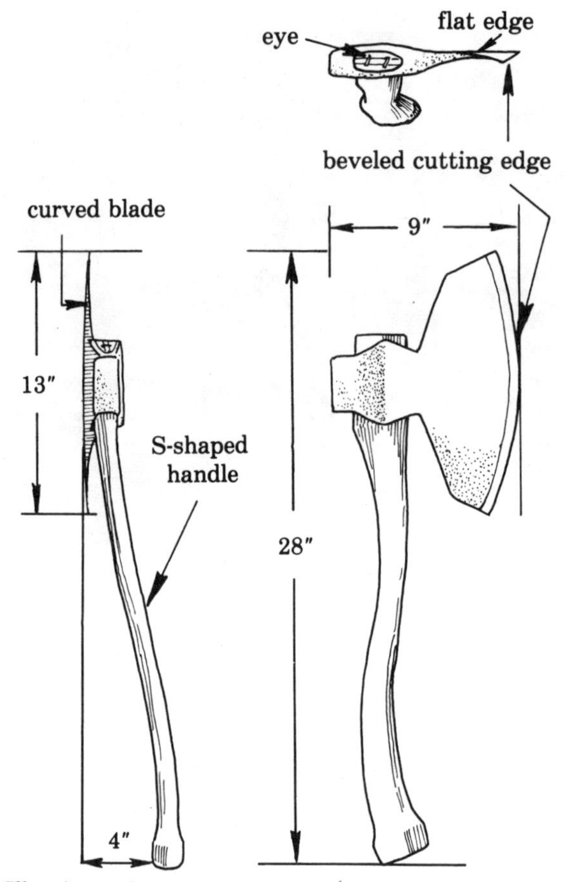

Illus. 7.1. *Three views of a broadaxe. Dimensions and shape of the head can vary considerably. The handle should bend away from the axe head to avoid scraped knuckles during hewing. To keep the corners of the axe from catching in the wood, both the blade and the cutting edge are curved.*

This arc, which may be ⅛ to ¼ inch, helps to keep the corners of the axe from digging into the log. The blade itself also curves, so that when the axe is resting on the middle of the cutting edge, there is a gap of ¼ to ½ inch between each corner of the edge and the surface it rests on. A low, wide broadaxe is easier to control than a tall axe, which tends to be less stable.

The head of a typical broadaxe weighs 6 to 12 pounds and the cutting edge measures 8 to 14 inches from corner to corner. With a wide blade, you can hew long, nicely feathered shavings. A broadaxe handle can be thicker and less smooth than that of a splitting axe. When chopping wood, you need a smooth finish so your hands slide into position, and the wood handle is narrow to whip the head for added power. In hewing, though, both hands remain in one position, so the premium is on a comfortable, rigid handle. The handle length can vary from less than 1 foot for a hewing hatchet to 4 feet for a Japanese broadaxe. About 28 inches is a good size for an ordinary broadaxe, but it is really a matter of personal taste. The handle of a broadaxe bends away from the flat side of the blade, so that the user's knuckles aren't scraped against the log during use. You can make a good broadaxe handle from straight-grained hickory, birch, maple, ash or oak.

BASIC HEWING METHODS

There are many ways to hew a log. Some people set the log on the ground and hew mainly by eye. A Swedish friend of ours, Wille Sundqvist, tells of an old master axeman who was famous for his logwork when Wille was a small boy. This man stretched a string from one end of the log to the other, right above the plane he intended to hew. He would then hew to the string, never cutting it. The most common hewing method, however, uses snapped chalk lines. First the log is set on a base of notched cross logs and chalk lines are marked down the log along the hewing lines. The log is then turned over and lines are also marked on the other side. The person wielding the axe then scores (notches) the log across the grain, cutting nearly to the depth of the chalk lines, and then cuts free the bulk of the waste wood between the notches. After this roughing-out stage, the hewer rescores the log, sets it on a pair of trestles, and carefully hews it to the final dimensions, working first along one chalk line, then the other.

Most hewers move forward as they work, but some walk backward. The shape of an old axe handle may indicate how the axe was used. Most often, the handle was bent away from the flat side of the broadaxe when the hewer worked forward, while a short, straight handle was more often used for hewing backward with the log on the ground. I've also seen a photo of a man standing on the log he was hewing; this required a handle bent toward the flat side.

BARKING

Some hewers work with the bark on the logs, though it's more common to remove the bark first. Of my two logbuilding teachers, Daniel O'Hagan and Peter Gott, Daniel likes to leave the bark on, while Peter and I remove it. When Daniel scores the sides in the old way, with a long-handled axe, he stands on the log, legs apart, chopping out chips from two directions. The bark aids his footing. He also feels that the bark gives the broadaxe something to bite into. Logbuilders who work with debarked logs generally feel that bark contains grit picked up while skidding logs to the work site, which dulls the axe blade. Also, a chalk line on bark is harder to see than on a smooth, clean log. Chunks of bark sometimes break away, leaving no line at all to follow. For starting out, at least, it's probably better to bark the logs before you begin hewing.

Barking results in huge piles of waste. Dry bark burns nicely, and makes great garden mulch, but home shredder/grinders cannot handle the quantities that will quickly accumulate. Try to plan how you'll dispose of the stuff before you get started, or you will end up with mounds of it, interfering with the rest of the operation.

Use a single- or double-bit axe to chop a strip of bark 3 to 5 inches wide, the full length of the log. Chop on the far side of the log from where you stand. Then rotate the log 180 degrees and chop a second strip. Remove the bark with a tool called a barking spud, or use a garden spade. The bark

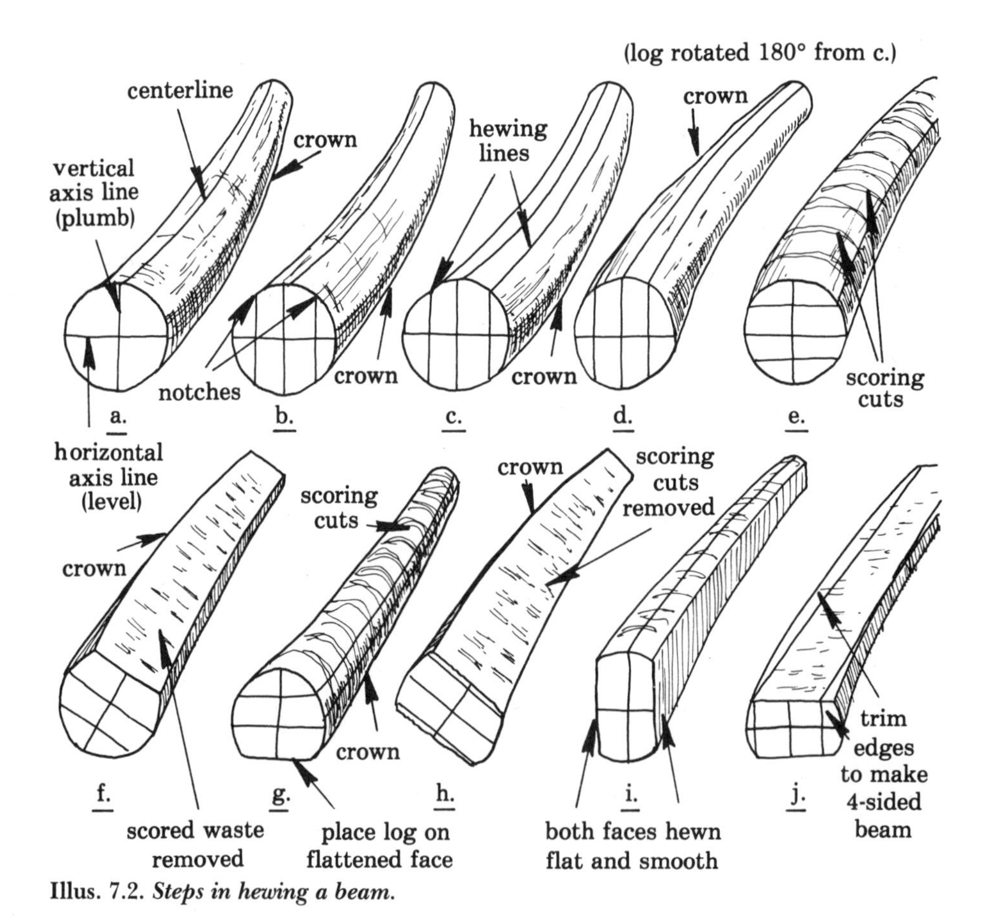

Illus. 7.2. *Steps in hewing a beam.*

Photo 7.2 ***Barking starts with an axe.***

from logs cut in spring and early summer often lifts off in large slabs. Bark that resists peeling must be chopped off, or removed with a large drawknife. Protruding limb stubs from rough bucking should be axed or sawed flush with the main trunk.

LAYOUT

To hew a log properly, you need good chalk lines to work from. To be sure that the lines are aligned properly, plumb a set of vertical lines at each end of the log and snap the hewing lines between them. This way, you can ensure that the hewing marks will be parallel to each other and properly centered on the log. You may find a two-color system useful, marking an initial set of orienting lines in blue and the cut lines in red, or you can do the whole thing in a single color.

Photo 7.3 ***Using a barking spud.***

Place the log to be hewn on a pair of short cross logs, sometimes called cribs. The cribs are about 3 feet long and 6 to 10 inches in diameter. Chop a rough saddle notch in the middle of each crib.

Support the log at both ends by tacking a short piece of wood from the log to the crib. Freshly peeled logs are often wet and slick. They may slide downhill, or rotate unexpectedly.

Recut the ends if the log is dirty, or if the original bucking cuts are not square. Remeasure the log to be sure there are at least 2 to 6 inches of extra length to allow the final trim during notching. Rotate the log swag (the crown) down. Sight along the log from both ends. You can see the full contour better with the swag down than you can with the swag up. Use a plumbed level to help align the log. Proper alignment is necessary if both sides are to be hewn evenly. Crooked or twisted logs often require a degree of subjective judgment.

Photo 7.4 *Barking summer-cut yellow poplar.*

Photo 7.5 *Using a timber carrier to lift a barked log onto cross cribs.*

Photo 7.6 *Snapping the centerline.*

Photo 7.7 *Locating the vertical (plumb) centerline.*

Photo 7.8 *Locating the horizontal (level) centerline.*

Photo 7.9 ***Cutting a small notch to hold the chalk line in place.***

Stretch a chalk line down the middle of the log. Pin the line with awls inserted in the center of each end of the log. Chalk lines snapped on a cylinder can result in curves that do not follow the desired hewing plane. This easily happens when the log is bowed or the working site is not level. Use a plumbed level set in the middle of the log as a guide for lifting the chalked string to ensure that the line is directly above the log. Hold the level plumb, about ½ inch from the stretched string at the approximate midpoint of the log. Raise and release the string parallel to the level. If the top and bottom centerlines and hewing lines are all done in this manner, the resulting lines will be parallel.

Next, mark orienting lines at both ends of the log. Use the level and a pencil to draw a vertical (plumb) centerline down from the center chalk line you just snapped along the log. Then draw a horizontal (level) line, somewhere near the center of the log (the exact position does not matter). Measure an equal distance to the right and left of the vertical, along the horizontal line, to locate the hewing lines. For a 6-inch beam width, measure 3 inches on each side of the vertical centerline. Use the level to draw vertical lines indicating the width of the hewn log. Cut a small V-notch where each new line meets the perimeter of the log end. Repeat this procedure at the other end of the log.

Photo 7.10 ***Drawing vertical lines indicating the width of the hewn log.***

Photo 7.11 ***Snapping the hewing lines.***

Snap two chalk lines along the top of the log to indicate the hewing plane for both cuts. Use the plumbed level to keep the chalk lines parallel. Rotate the log exactly 180 degrees (crown up), and use a level and the penciled lines on the log ends to make sure of the exact positioning. Snap two hewing lines along this side of the log. If you can cut to the plane indicated by these four cut lines, the sides of the log will be exactly flat and parallel.

SCORING

Scoring is the process of cutting across wood fibers at regular intervals, outside the cut lines, so that chunks of wood can easily be

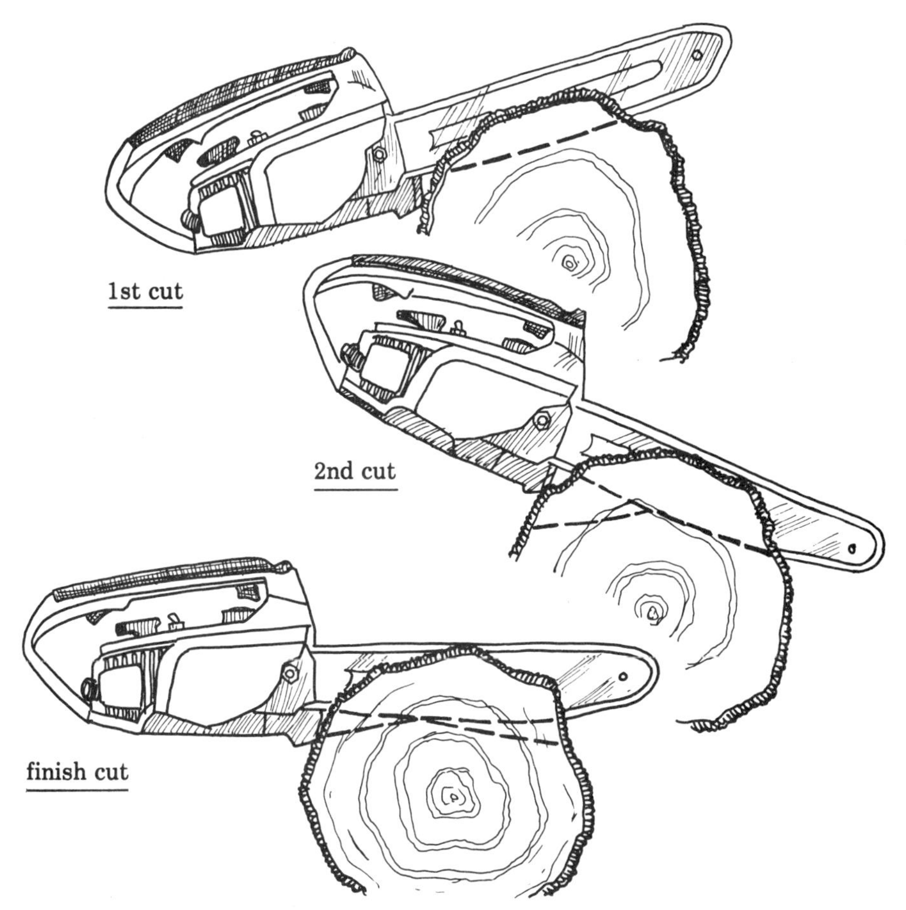

Illus. 7.3. ***Scoring a log with a chain saw. First cut the near side with the blade angled upward, then pivot the saw to cut the far side. Finish with a horizontal cut. Always keep the blade tip exposed to avoid kickback.***

chopped off the side of the log. Expert axemen can hew some kinds of wood without scoring, but this is exceptional. With accurate layout and scoring, you can hew exactly to the scoring cuts for the final surfacing. Traditionally, V-cuts for scoring were chopped at regular intervals with a thin-bladed, double-bit axe, though a single-bit axe will do as good a job with less risk. You can also score with a crosscut saw or a chain saw. For crosscut or chain saw scoring, rotate the log 90 degrees so the hewing lines at the end of the log are horizontal. With a chain saw, angle the blade up slightly, and stop the cut just above the line on your side of the log. Then finish the cut by angling the tip of the blade down, cutting to the line on the opposite side of the log. Beginners may want to stop these cuts 1/8 to 1/4 inch above the lines, but scoring to the lines is best. On large-diameter logs, the

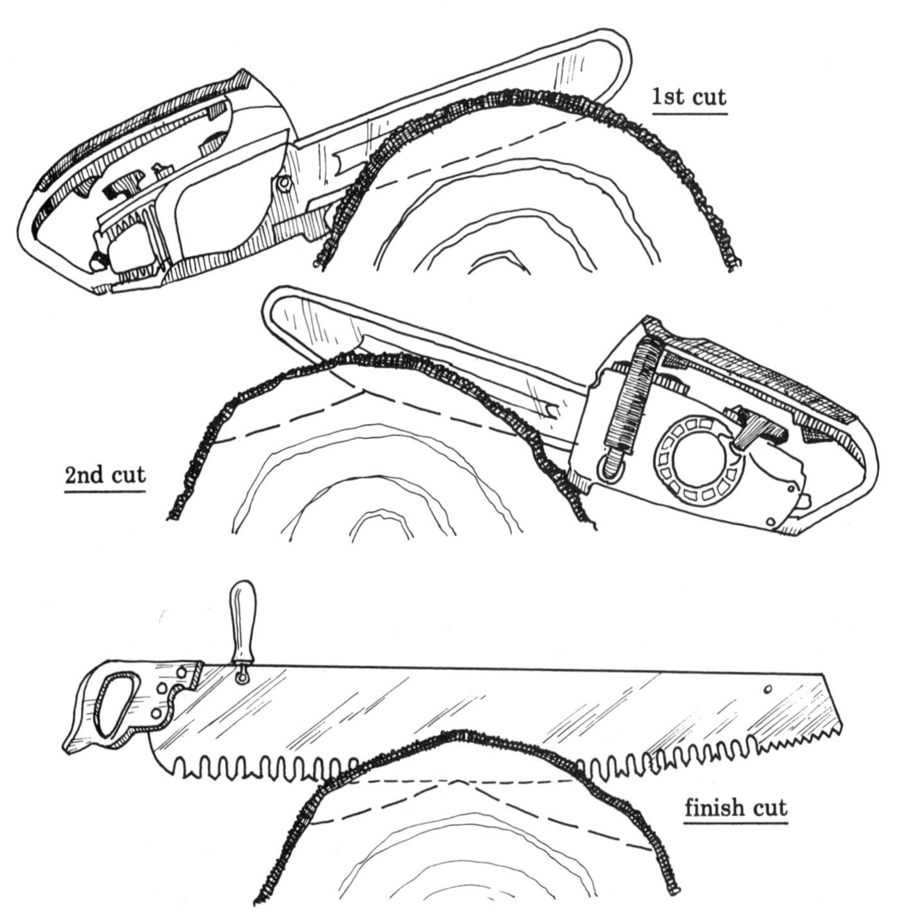

Illus. 7.4. ***Scoring a large-diameter log. Cut from both sides with a chain saw; finish with a hand saw held horizontally.***

Photo 7.12 *Scoring. Face guard protects eyes from flying sawdust. Ear protectors are also recommended.*

Photo 7.13 *Splitting off scored pieces from the log end.*

Photo 7.14 *"Juggling." Note the tilted angle of the log.*

Photo 7.15 *Rescoring.*

curvature of the chain saw bar can cause a concave kerf. To avoid this, saw from each side in succession, keeping the saw angled up a little at the center of the kerf. Finish with a hand saw to cut the kerf flat.

Space the scoring cuts 5 to 10 inches apart. Easy, straight-splitting logs can be scored at greater intervals, up to 2 or 3 feet. But space the kerfs as close as 1 inch apart around knots and areas of irregular grain. Several kerfs spaced 3 inches apart at each end of the log will ensure that you can start chopping the notched sections cleanly.

The scored pieces, sometimes called juggles, are chopped off with a polled axe. Rotate the log about 30 degrees from horizontal, to a comfortable angle for chopping the far side (the base of the scoring kerfs should tilt away from you). Place the axe blade against the log end, ¼ inch outside the hewing lines. Strike the poll (the flat end of the axe head) with a wooden club to remove the first hunk of wood, then begin chopping them off with the axe. Try to keep the cuts ⅛ to ¼ inch above the base of the saw kerfs.

After knocking off the juggles, lightly rescore the log with the axe, cutting slightly deeper than the finished hewing plane. Space the cuts 3 or 4 inches apart, in a regular pattern. Axe scoring severs the wood fibers, making hewing considerably easier and cleaner. Neat scoring can result in an attractive, decorative effect on the finished log, but the scoring should be shallow; deep cuts will hold moisture. When you've rescored one side of the log, reverse it and repeat the process on the other side, chopping off the juggles, then rescoring.

HEWING

Hewing is actually removing the last wood down to the hewing line. Although a broadaxe makes the work easier, any sharp axe can be used to hew a log. For notching and leveling floor joists, a small, short-handled adz is very useful. We have several older neighbors who still hew locust barn sills with a standard double-bit axe, placing the log just above the ground on cribs, although this method requires the hewer to bend over.

I hew with the log on a pair of trestles—rough, heavy, sawhorses consisting of a half-log bench and four hardwood legs. My trestles are 30 inches high and about 24 inches long. I like short trestles because I can hew on either side of the log without changing its position. I use a soft, light wood, such as pine, for the bench and well-seasoned locust heartwood for strong, durable legs. It's nice to have several sets of these trestles around the work site.

To hold the axe properly, grip the handle with your right hand (if you are right-handed) just behind the axe head. Hold your thumb forward, not around the handle. Place your left hand well back, at a point of comfortable balance. Brace your left forearm against your leg or hip. If you

Photo 7.16 ***Hewing.***

are left-handed, the procedure is just the opposite. You will tire less if you maintain good balance, a regular working rhythm, and develop the sense of relaxation that comes with practice. To get down to the proper working height, bend both legs into a sort of fencer's position. Keep your back in line with your center of gravity. This position is difficult at first. As a learner, I could hew only 5 or 6 feet before needing a rest. I now easily hew both sides of a 20-foot log with pauses only to move braces. A sharp, properly shaped axe also makes a big difference.

Photo 7.17 *Positioning the axe slightly askew to make a beveled cut. Note position of the thumb.*

Photo 7.18 *Turning a log with a small cant hook.*

Photo 7.19 *Hewing thin shavings.*

To begin hewing, position yourself to sight along the log from one end. Angle the axe blade away from the chalked hewing line slightly and carefully hew down the vertical line at the end of the log, attempting to split the line. Stop about two-thirds of the way down, then carefully hew along the chalk line on the top of the log for 3 to 4 feet. Tilt the axe slightly off plumb, as for the end cut. Return to the end, straighten (plumb) the axe, and begin hewing along the side. Make successive forward passes with the blade as you work your way along the length of the log, hewing with a circular, rhythmic motion always in the same plane. Some axemen hew straight down to the bottom of the log, but this risks ripping the edge fibers and increases chances of hewing away from correct plane. Hewing the lower third requires bending over, and can put a strain on your back as you hew through the bottom. There is also more danger of an accident. After hewing about two-thirds of the way through on both sides, I turn the log over and hew the remaining section.

Learn to hew wide, fine wafers. Concentrate your attention between the log and the inner face of the axe. Cut with the midsection of the axe blade, so corners don't nick or stick in the log. Try to achieve a flat surface.

Knots may require hewing from two directions to prevent the wood fibers from tearing unevenly. Closely spaced scoring with saw or axe helps considerably in keeping such cuts smooth. I sometimes stand on a pair of

Photo 7.20 ***Joists hewn square on four sides.***

extra trestles and hew knots backward with a polled axe. In some cases, the grain of the wood twists into the log, causing the blade to sink deeply and tear loose roughly. The best solution is to be ambidextrous and have a left-handed broadaxe. You can then hew from the other end so that the grain harmlessly runs outward.

Four-sided beams are hewn like two-sided logs, but with two extra finish cuts. Hew the wide dimension first, then square off the top and bottom. Use the original crossed center lines at the log ends for laying out the extra two hewing planes. Three-sided logs are sometimes required in logbuilding to keep the first course of logs above squared sill logs level. Hewing these requires marking out only one additional plane.

POSITIONING THE LOGS

Lift heavy logs onto the trestles carefully, and have help. Lift with bent legs and a straight back. Really heavy logs can be slid up an improvised ramp, or lifted with shear legs (see illus. 9.2).

Hold logs in position with hewing dogs, which are large, forged-iron staples pounded into the log and the trestle. Some axemen flatten the curved portion of the log above the trestles to provide more secure footing

Photo 7.21 ***Scrap wood "staples" secure the log to the trestles.***

for the log. You can also wrap a chain around both the trestle and the log, and tighten it with a lever-and-cam device known as a chain load binder. Heavy logs tend to stay put, while short, light logs are harder to secure.

A friend, John Chiarito, devised a simple holding system that I like to use. His variation of the traditional hewing dog utilizes scrap 1 × 1s nailed to both log and trestle. These are inexpensive, variable in size, and unlikely to damage the axe blade. I call these fasteners "staples."

Once the log is on the trestles, set it plumb. This usually requires temporary wedging with chips, which are always plentiful around the hewing area. Nail two diagonal staples at the far end of the log and one staple at the starting end, on the side you aren't hewing. If the log is still unstable (from irregular curvature) flatten the bottom or nail temporary holding cleats to the trestle bench top. After hewing past the trestle, nail a

Photo 7.22 *Bending a broadaxe handle.*

Photo 7.23 *Hewn logs in storage.*

fourth staple at the starting end on the hewing side. Remove the staple at the far end as you approach, then replace it when you turn around to begin hewing the second side.

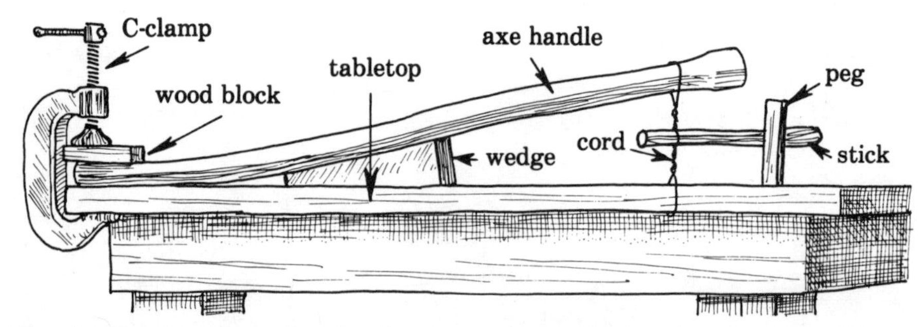

Illus. 7.5. ***Bending a broadaxe handle. Clamp the axe-head end, wedge the middle, and tighten the other end with a stick twisted in heavy cord or light-gauge wire. The peg mounted to the tabletop keeps the stick from unwinding.***

MAINTAINING AXES

Good, new broadaxes seem to be unavailable, unless you can afford to hire a blacksmith to do a custom job. Old broadaxes, however, can be found at junk shops, flea markets and antique tool dealers, usually at prices from about $25 to $100, depending more on who is selling than on the condition of the tool. Rusty, slightly pitted or chipped axes can often be refurbished, though the cutting edge of old axes may be too hard to be sharpened with a file or hand stone, in which case retempering may be called for. Examine the steel blade; often, someone has beveled the inner face, perhaps ruining the tool for hewing. To use such an axe, you must remove this inner bevel, and this could result in losing the steel cutting edge, since axe heads were traditionally forged from soft iron, with a high-carbon cutting edge forge-welded to the head. Consult a blacksmith with experience in toolmaking if you are interested in obtaining an axe in this condition. If it is salvageable, the blacksmith may also be able to put a curve across the back side, or retemper the blade for sharpening.

Chips and pits can sometimes be removed with an electric emery wheel. Use a face shield, and be very careful not to overheat the cutting edge and destroy its temper (a condition usually indicated by a blueing of the steel). To avoid this problem the axe head should be hand-held while being ground and immersed in a bucket of water as soon as the metal gets hot.

Sharpening begins by shaping the cutting edge with a grinder or file. Electric grinding is risky, because it's easy to overdo it. If the edge is in fair condition, use a fine mill file for most of the work, one about 10 or 12 inches long, with a handle and a guard at the butt of the tang (where the file sticks into the handle). The guard, which may be a piece of copper, is important because a slight slip of the file can result in sliced fingers. Leather guards are not effective.

For filing, secure the axe head to the edge of a workbench with C-clamps, or use a vise if the axe has the handle attached. Sometimes a shaving horse will also work. The inner face of the axe blade must be flat along a vertical axis from the cutting edge to the eye. Grind or file off any bumps or protrusions. The flat side should be smooth so that friction is minimized as the blade slides past the log.

The angle of the cutting edge should measure 22 to 25 degrees. Too wide a bevel makes cutting difficult. A narrow angle requires frequent sharpening and is more likely to chip or nick. File until you cannot see any light reflected off the cutting edge. (Files cut on the forward stroke only; pressure on the return stroke will dull the file.) You should now feel a slight burr along the entire back side of the blade. To feel the burr, draw your finger away from the edge. Never run your finger along a sharp edge.

There is a great deal of controversy among woodworkers regarding the best shape for cutting edges. Some axemen advocate a slightly curved bevel, while others believe that the bevel should be shaped flat and kept that way. A slight curve at the edge sharpens faster, and the shape helps lift off the chip, but the angle changes slightly with each sharpening, periodically requiring regrinding. I prefer to maintain a flat bevel at a constant angle.

After filing, dress (whet) the edge with a hand stone. Use a round Carborundum stone made for axe sharpening. The stone is sold in a combination grit, but I seldom use the coarse side; a file cuts faster. For sharpening oil, use a 1:1 mix of 30-weight motor oil and kerosene. Use the stone to take off the burr on the flat side. Work in a circular motion along the edge, keeping the stone flat to the face to prevent a bevel from forming on the inner face. Stop when you can feel a slight burr along the entire bevel side. Dress the bevel side in the same manner, carefully maintaining the flat bevel made by the file. Again, stop when a burr is felt on the flat side. Gently remove the back burr, leaving a very slight burr along the bevel. Alternately dress each side, always with lighter strokes, until all traces of a burr disappear. When your broadaxe is razor sharp, make a guard to protect the edge and yourself when the axe is not in use. The protector can be leather, a slit hose or a piece of folded, taped cardboard.

I have never seen a broadaxe for sale that had a usable handle. You can buy an axe secondhand, but you'll have to put on the handle yourself. Most broadaxe heads are symmetrical, so they can be made for left- or right-handed use, according to the way you insert the handle. For forward hewing with the log elevated on trestles, the handle must take a severe bend just behind the head, for knuckle clearance. The handle should take a second, gradual bend at about midsection, so the overall shape is a rough S. A properly shaped broadaxe handle will align both of the hewer's hands with the axe head, and the axe head will hang vertically when the axe is loosely gripped.

Handle blanks should be split from a long length of wood oriented so that the annual rings will cross the eye of the axe head, when viewed from above. Hold the blank in a shaving horse and shape the contours with a drawknife and spokeshave.

Handles should be well dried before fitting to the axe head. Put the handle stock in a warm place (such as behind a stove) for a few days. Fit the handle to the axe head and secure it with wedges, driven into a 2¼-inch slot cut in the handle end, but leave the excess extending above the axe head. Immerse the fitted handle in boiling water for 20 or 30 minutes, then quickly clamp the head with a C-clamp to a stout board or bench top. Working while the wood is hot, bend the handle away from the board about 30 degrees. Insert an improvised wedge just behind the head, then pull the end of the handle back toward the board, so that it takes on an S-shaped curve. Wire the end down, or fasten a loop of rope around the handle and the board, with a stick inserted and twisted to provide tension. Leave the handle in the jig for about one week. Check the set by testing for spring at the C-clamps or toggle rope. When the handle doesn't give, the final bend will be permanent. Axes should be kept in a nonheated area, such as a garage. The dry air in a heated building will cause the handles to shrink and loosen.

CHAPTER VIII

NOTCHING

Hewing is half the job of preparing a log, and notching is the other half. It takes just about as long to notch a log as it does to hew it. Careful notching is worth the time it takes because the notches are what really hold the building together. Properly made corner notches should support 70 to 80 percent of the weight of a log building. Corners can loosen if the blocking and framing carry too much load. Loose corners also allow moisture penetration and air infiltration.

The most visible notches on a house are at the corners, but other notches usually occur on floor joists, which are generally notched into the sill logs or the foundation wall, and on some types of partition walls, which tie into notches in the exterior walls. Notches are also required in wall logs that form lintels at window or door openings, and on the ends of the rafters (as explained in chapter 10, Roofing). Another form of notch—a splice—joins two or more short logs into a longer piece.

CORNER NOTCHES

Simple corner notches, which are generally used with round-log construction, consist of a single cutout segment, usually on the lower half of the log. The depth is half the log diameter, allowing each succeeding course of logs to rest directly on the course below. The most common example of a corner notch is the saddle notch, which is curved to match the shape of the log below it. It is fairly easy to hew out this notch with only an axe, but difficult to get a perfect fit. Although saddle notches are quite adequate for log construction—and quite popular among logbuilders—perfect fits must be obtained nevertheless, since the geometry of these notches allows water to seep between the logs and become trapped there if the logs do not fit together properly.

Lapped notches are made by cutting away the upper portion of a lower log and the lower portion of the upper log. A variation of the lapped notch (used on toy Lincoln Logs) is called the lock notch. In it, one-quarter of the log height is removed from the top and one-quarter from the bottom, and the middle of the log is left solid. When this type of notch is squared off at the log ends and used with no extension of the log beyond the notch, it's called a square notch. The result is a tidy corner, but there's no provision to stop logs from sliding out, unless they're spiked.

When a log has an upper and lower notch (as in the Lincoln Log type), it is referred to as having a compound notch. Angled compound notches have the added complexity of being sloped, rather than horizontal, along their notch faces. Angled notches are harder to cut, but they lock log corners more securely and tend to shed water better than the simpler types. Two excellent angled compound notches are the dovetail notch and the V-notch.

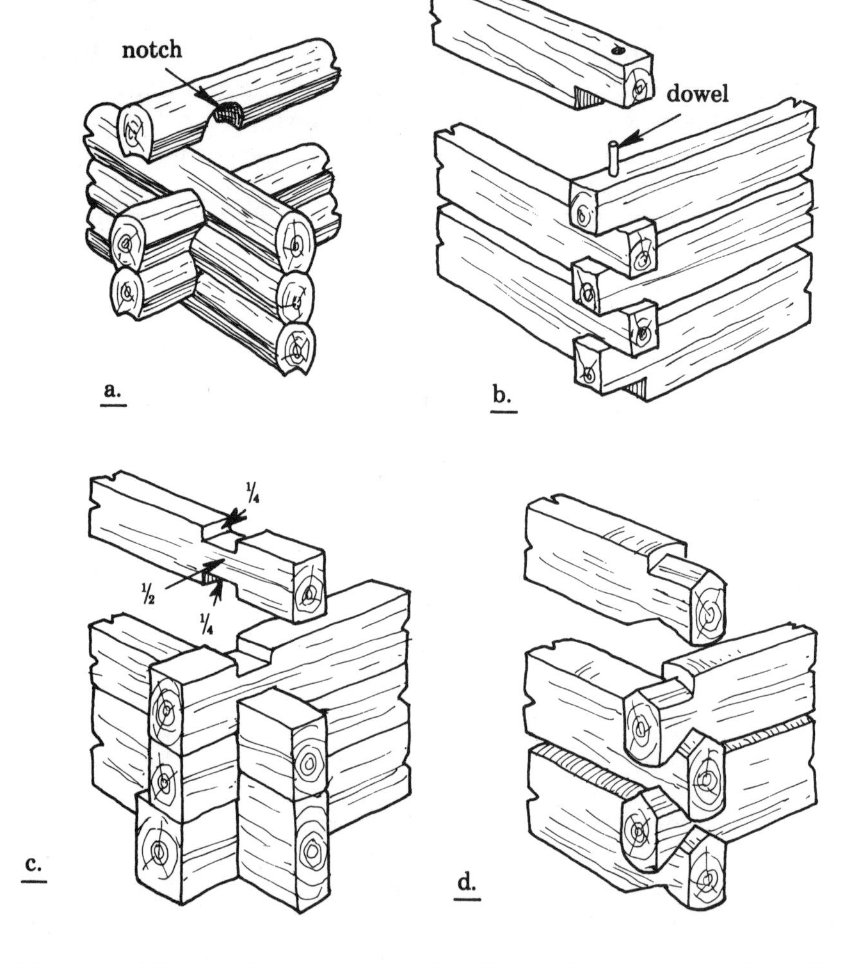

Illus. 8.1. ***Different styles of corner notches (here and on facing page). (a) Saddle or round notches. (b) Half-lapped notches secured by dowels through each notch. (c) Lock notches used with Swiss-style logs squared on four sides. Note extra width at sill log and height of first cross log. (d) Half-dovetail (or common dovetail) notches. (e) Full dovetail notches have compound angles. (f) V-notches.***

The dovetail notch is an adaptation of the lapped notch, but with the notch faces sloped to help pull the log into place. The V-notch has an upper and lower notch (compound), each with two opposing faces, like a gable roof. These upper and lower notch angles are at 90 degrees to each other, so the logs lock when stacked (the effect is of two gable roofs set one on top of the other, with their ridge lines at right angles).

The dovetail notch has a reputation for being complex and difficult to master. However, this notch is not hard to lay out and cut, once you understand how it works. The common dovetail notch, which I use, is sometimes called a half-dovetail because the bearing face of each cutout is cut along one plane only. A full dovetail notch is one cut with bearing faces that slope along two planes at once, producing a notch that is self-tightening. Half-dovetails can move during construction, but are perma-

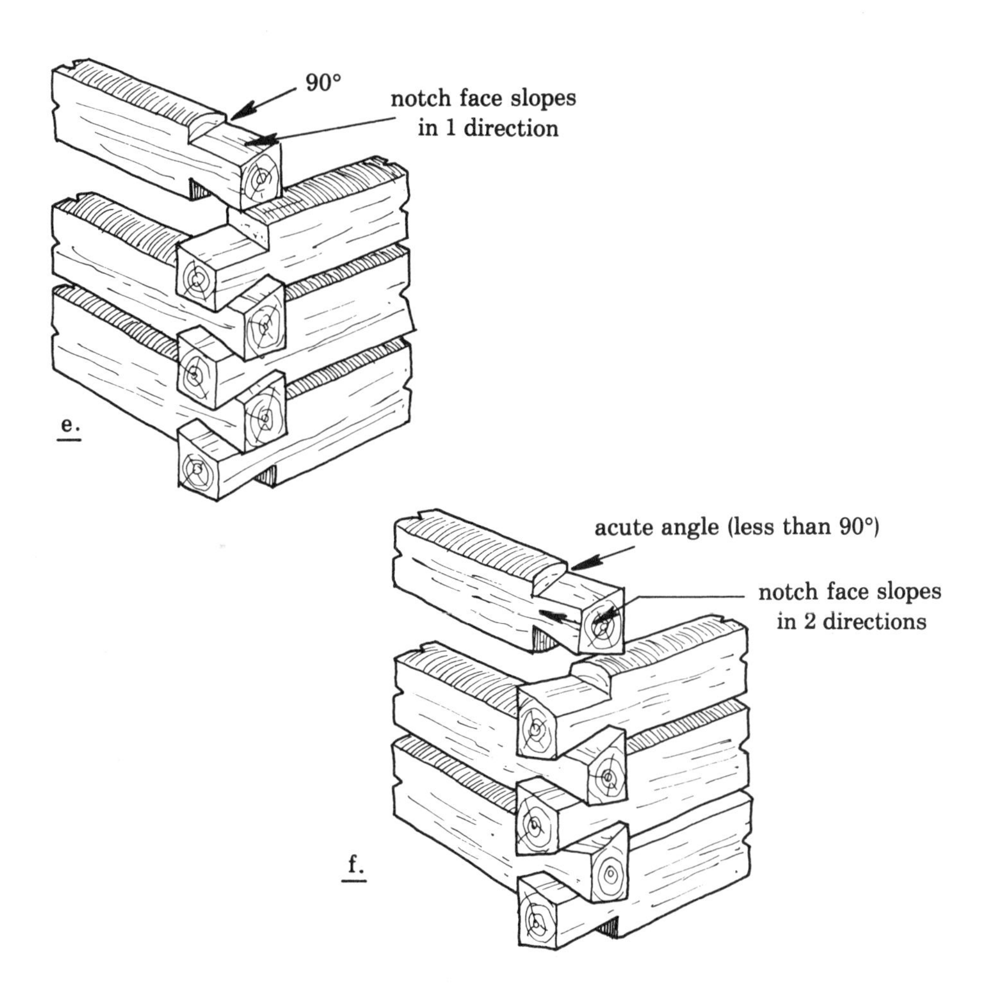

Photo 8.1 ***Half-dovetail notches.***

nently locked once a cross log is placed on top. Plate logs—which have no cross logs—are sometimes notched with full dovetails to withstand the outward thrust of the roof rafters.

The half-dovetail can be laid out and cut in less time than a V-notch, and gaps caused by log shrinkage can often be closed with a sledgehammer. The main disadvantage of all dovetail notches, however, is that they include a triangular section of sapwood which can break off, or deteriorate faster than the rest of the joint. Also, increasing the gap between logs in order to build into them more chinking thickness requires decreasing the slope angle of the notch face (otherwise, the tip of the upper corner may not be wide enough to notch), and this, of course, lessens the locking ability of the notch.

Like the half-dovetail, the V-notch requires a cross log above it to hold it in place. The main advantage of a V-notch (compared to a dovetail) is that considerably more sapwood is removed from the upper notch, yielding a more durable joint. The V-notch is more time-consuming to lay out and cut: Four planes must be aligned, and the ridge line of the upper

Photo 8.2 ***V-notches; the Gott homestead.***

notch can cause problems if the log is warped or is hewn off-square. If there is crossgrain shrinkage, the V-notch cannot be tightened. For these reasons I prefer the dovetail, though many logbuilders do beautiful work with V-notches.

LAYING OUT HALF-DOVETAIL NOTCHES

One way to cut notches is to rough them out by eye when the logs are on the ground, then fit them together more accurately using scribing tools or trial-and-error adjustments after the logs have been raised into place. When Peter Gott began logbuilding, he wanted a method that would allow accurate notches to be laid out while the logs were still on the hewing trestles, minimizing the adjustments necessary after the logs had been raised. The system he developed is based on first snapping a chalk line along the middle of the log, from end to end, and using that as a basis for laying out and cutting the notches. Measuring notch depth from the centerline depth is a major advance in logbuilding technique because it eliminates problems caused by log taper. Logs are always stacked so that butts and tips alternate on the same wall. As long as the centerlines marked along the lengths of the hewn faces are kept level, all the logs on the wall will be horizontal.

In any log building without chinking, the amount of wood removed from the stack of logs that forms one corner accounts for one-half the total height of the wall. (To put it another way, if the same logs are not notched at all but simply placed one on top of another they will form a wall twice as high as the wall of notched logs.) It follows from this that the depth to which wood is removed from one end of any of these logs to form an individual corner notch equals half the height of that log. Dovetails are compound notches and have two cutouts per log, one above the centerline and one below. Thus, the depth to which each of these two cutouts must be made has to equal one-quarter of the log's height. Using Gott's method of measuring from an arbitrary but carefully located centerline outward toward the uneven edges of a log, rather than inward from those edges, allows measurements for dovetails and other complex notches to be made with complete precision.

Begin by laying out a new horizontal line along the length of what you select as the exterior hewn face of the log you wish to notch (the side of the log that will be on the outside of the building). Use a chalk line as you did when marking out the log for hewing. Lay out this line by eye, as shown in photo 8.3. Remember, since all notch measurements are based on this centerline, there need be no preexisting relationship of this line to either of the hewn logs' outside edges. The centerline is the arbitrary base of all subsequent measurement; theoretically it doesn't even need to bisect the width of the log at all, but if it does, you will be allowed a maximum amount of wood to work with on either side of the line.

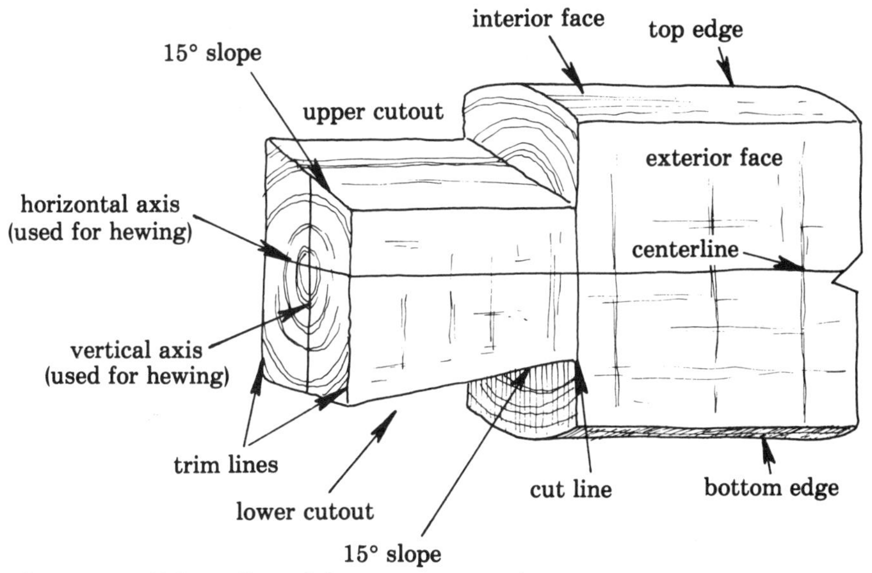

Illus. 8.2. *Half-dovetail notch layout terms.*

Next, mark out the trim length of the log; that is, the length the log will be after all notching cuts are made. The trim length consists of the interior room length, plus the thickness of two wall logs (one crossing each end of the log being notched), plus two notch extensions. Notch extensions are additional amounts of wood left on the log ends. They project beyond the point at which two notched logs intersect. Since end grain deterioration of logs is inevitable, notch extensions—which are actually extra allowances of end grain—protect the notches themselves from damage. I allow 3-inch extensions at each end of the logs I notch.

After adding together all these measurements, determine that the log is at least 1 inch longer than the trim length. If you wish to build, say, a wall with a 20-foot interior dimension, and you are using 6-inch-thick cross logs and allowing 3-inch notch extensions, the trim length of the log to be notched is 21 feet 6 inches, and the log itself should measure at least 21 feet 7 inches overall, before you start. Measure the trim length onto the log face, and mark it by drawing trim lines across the log face, perpendicular to the centerline, at each end of the log. Use a framing square for accuracy.

Now measure from each trim line toward the center of the log a distance equal to one notch extension and one cross log thickness. With 6-inch-thick cross logs and 3-inch notch extensions this means measuring in a distance of 9 inches from the trim lines on each end. Mark these points with a square as shown in photo 8.4, with lines perpendicular to the

Photo 8.3 ***Aligning the centerline by eye on the outside face of the log.***

centerline just as you marked the trim lines. These lines are called the cut lines and indicate where the notches stop and the wall log proper begins. The distance between the two cut lines should equal the interior dimension of the room you are planning to build.

With the cut lines and trim lines marked on one side, you are now ready to locate the positions of the sloping lines that form the notch

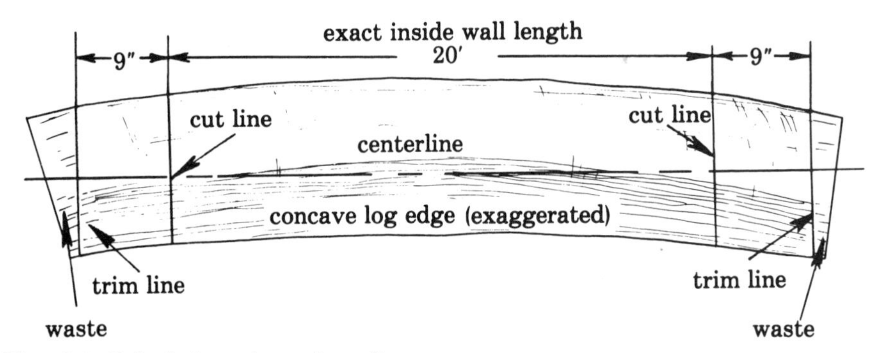

Illus. 8.3. *Calculating trim and cut lines.*

cutouts. The angle of each slope is the same for each log and is determined somewhat arbitrarily, as I will explain. The depths of the notch cutouts, however, vary with logs of different sizes. My way of determining the depths of the notch cutouts is by a simple formula which I have adapted from a formula originally worked out by Peter Gott. Before it can be applied, three additional measurements must be obtained: the height of the log (the distance between the two curved, unhewn faces which form the top and bottom surfaces of the log when it is in place), the vertical thickness (height) of the chinking above the log being notched, and the rise of the notch cutout itself. First find the height of the log: Measure it at a point midway along its length. Since the log inevitably tapers, this will give you an accurate average. Use a framing square and triangle (illus. 8.4) to obtain a measurement that includes the topmost point of the curvature on the upper face of the log, as well as the bottommost point on the curved lower face. The chinking thickness is something you have presumably decided upon during your initial construction planning (see chapter 11, Chinking). It should be a standard, constant measurement between all the logs, and thus may be considered merely an additional increment of log height. The chinking thickness I use is 1½ inches. In the formula, simply add the thickness of one band of chinking to the measurement you obtain for the height of the log itself.

The rise of the cutout, like the pitch of a roof, is the height to which the sloping face rises along the horizontal distance it travels. The rise is a function of both the width of the log (since the log's width is the horizontal distance the notch must cover) and the degree of steepness at which the slope rises. This degree of steepness, which I mentioned earlier as being somewhat arbitrary, is something to be decided at this time. Trial and error, plus some educated guessing, has led me to feel that, with a half-dovetail notch, a slope of 15 degrees or so is about optimal. Good locking together of the notches is achieved with this angle, and the amount of weak

Photo 8.4 ***Marking the trim lines and the cut lines with a framing square.***

wood at the outside corner of the notch is less than it would be if the angle were any steeper. By projecting an angle of 15 degrees over a horizontal distance of 6 inches (the width of the log being notched), a rise of about 1¾ inches is obtained.

Since the rise of each cutout on either side of the centerline is actually shared by two logs—the log being notched and the log that will lie across it (and presumably, the log being notched is also intended to lie across a similar log beneath it)—one-half of the amount of rise must be subtracted from each distance along the cut line above and below the centerline of the log being notched, because in each case that is the portion of the total rise (1¾ inches) that each cutout on the log being notched contributes to the complete half-dovetail, which is made up of two logs. The formula, then, for determining how deep to make the cutouts in half-dovetail notched logs is

$$\text{Depth of cutout} = \frac{\text{height of log} + \text{thickness of chinking}}{4} - \frac{\text{rise of notch cutouts}}{2}$$

As an example, let us say that the log being notched measures 12 inches in height. Using the same figures as above for chinking thickness and rise of notch cutouts, an actual cutout depth can be calculated as follows:

$$\text{Depth of cutout} = \frac{12'' + 1\tfrac{1}{2}''}{4} - \frac{1\tfrac{3}{4}''}{2} = 3\tfrac{3}{8}'' - \tfrac{7}{8}'' = 2\tfrac{1}{2}''$$

This is the distance away from the horizontal centerline—marked on the hewn faces as well as on the ends of the logs—that you locate the points which indicate how deeply you must cut into the log.

An enormous amount of time can be saved if you make a chart based on this formula that lists the depths of cut for a range of log heights. Then you can merely refer to the chart to find the correct distance to measure away from the centerline of the log you are notching instead of having to calculate the distance each time. A template incorporating the angles necessary to scribe the notch is also a useful time-saver and works better than the tool you would probably use otherwise, the sliding T-bevel. I use the template shown in the accompanying photos; it is made of roofing tin (you could make one of ⅜-inch plywood just as easily) and has a chart of cutout depths painted on one side. A diagram for making one of these templates is shown in illustration 8.5.

When you have calculated the depths of the cutouts for the log being notched, stand by the concave edge of the log (with its exterior face up, this edge will be the bottom edge of the log when it is in place) and mark this distance in two places on the cut line, once on either side of the centerline. Remember to measure these distances from the centerline outward toward the edges of the log. When you have marked the points, align the template or T-bevel so that its vertical edge is along the cut line nearest you (that is,

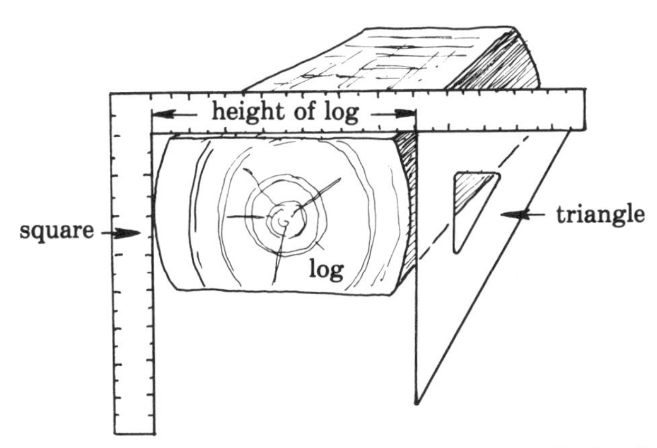

Illus. 8.4. ***How to measure log height using framing square and triangle.***

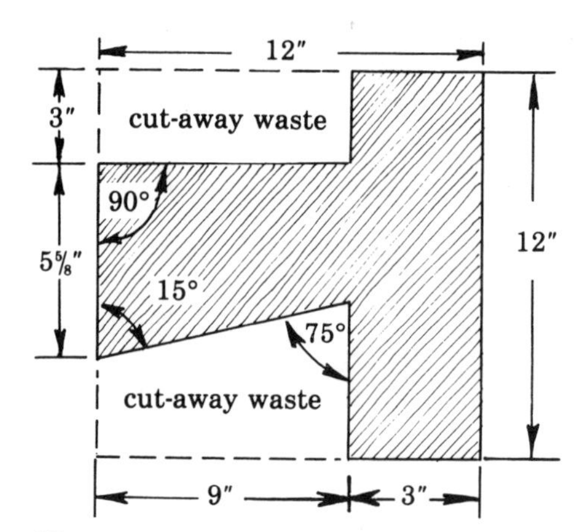

Illus. 8.5. *Pattern for dovetail template.*

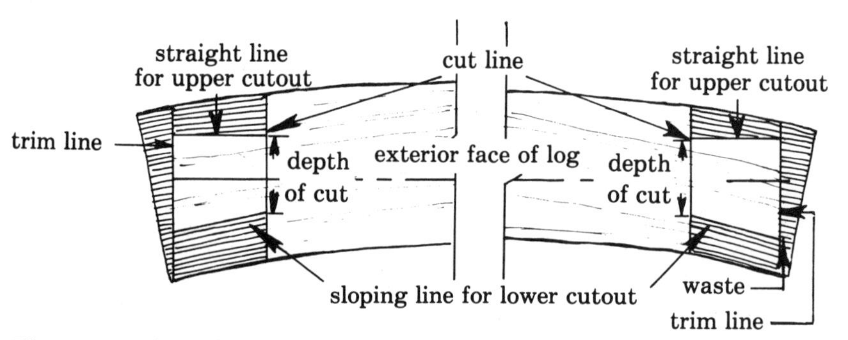

Illus. 8.6. *Scribing the notch cutouts.*

below the centerline), and its angled edge travels from the point you have marked outward toward the sloping line all the way from the cut line to the trim line, even if it means drawing on the curved portion of the log edge (photos 8.5 and 8.6). The triangle you have now drawn indicates the wood you must cut away to form the bottom notch cutout. Now slide the template away from you—still keeping its vertical edge aligned with the cut line—until the 90-degree edge of the template touches the point where the depth-of-cut mark on the far side of the centerline intersects the cut line, and draw a straight line along the template edge, parallel to the centerline, straight out to the trim line. This line indicates the lower edge of the upper notch cutout.

The next step is to extend the horizontal and vertical lines on the hewn log face all the way around the log. Begin by turning the log over on the trestles and marking out a new centerline along the log's as-yet unmarked interior face. This centerline must be exactly parallel with the original one on the opposite hewn face. Viewed from the end of the log, if the two centerlines were connected by the line across the log end, they should form a level plane dividing the log in two. To accurately place this second centerline, use dividers to mark the distance from the end of the first centerline to the horizontal axis line (used for hewing) which is probably still visible running across the ends of the log. Transfer this distance onto the interior hewn face in order to locate the point at which the centerline meets the end of the log exactly the same distance above or below the old axis line (see illus. 8.7). Do this at both ends and snap a chalk line between.

Dividers can also be used in conjunction with the framing square to lay out the new sets of trim and cut lines, should the log be warped enough to necessitate this extra treatment. To do this, lay the long leg of the square on the just-snapped new centerline, with the short leg hanging over the

(continued on page 136)

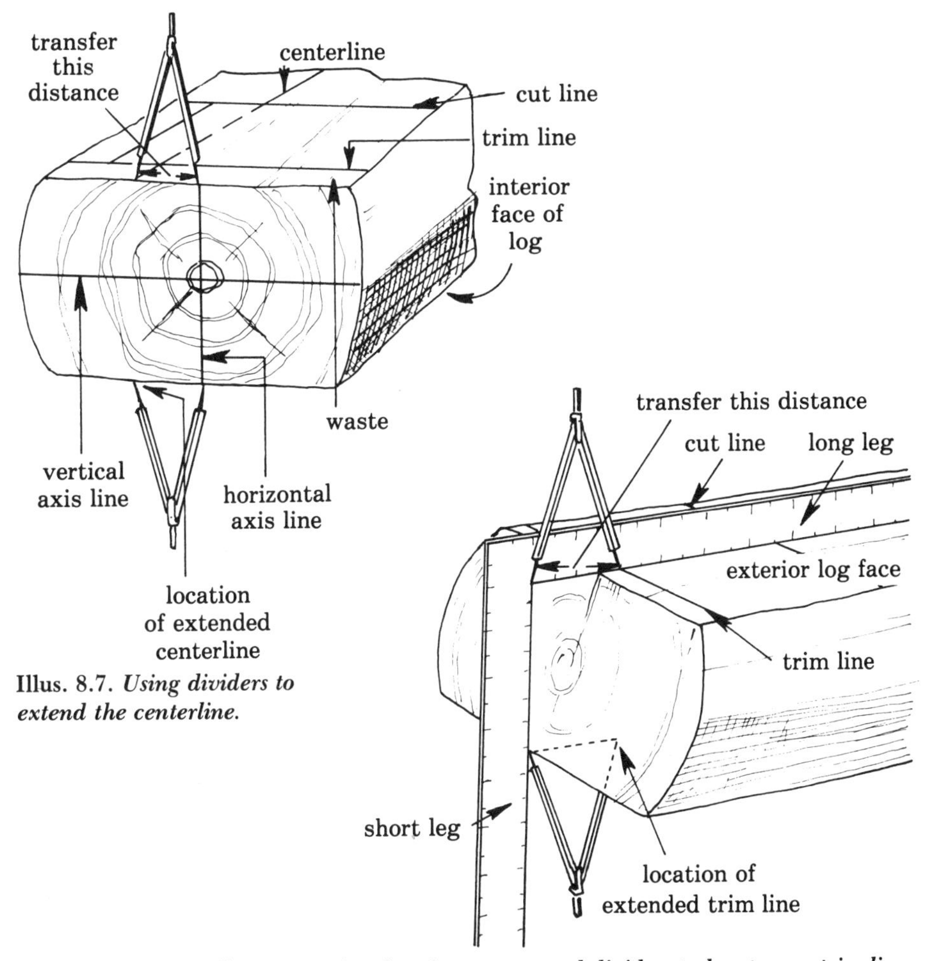

Illus. 8.7. ***Using dividers to extend the centerline.***

Illus. 8.8. ***Using framing square and dividers to locate new trim lines.***

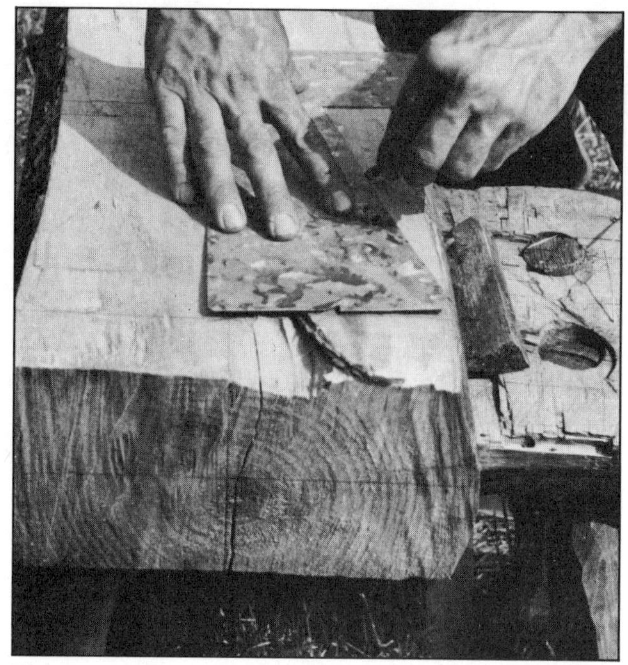

Photo 8.5 ***Using a homemade template to draw the angled cut for the bottom notch.***

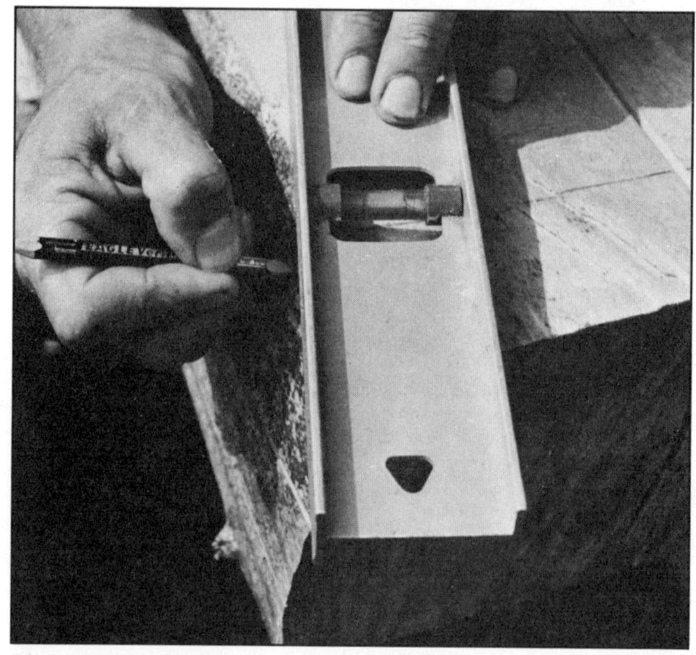

Photo 8.6 ***If the bottom notch terminates before the trim line, extend the angled line across the curved bottom of the log.***

Photo 8.7 *Using a flexible stainless steel ruler to draw lines across the top and bottom of the log.*

Photo 8.8 *Trimming the log ends. Use a hand saw to make a clean, accurate cut.*

Photo 8.9 ***Redrawing the horizontal centerline after the end is trimmed.***

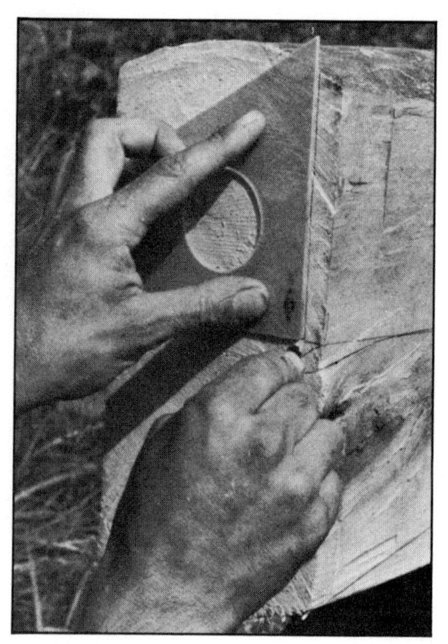

Photo 8.10 ***Drawing a vertical guideline using a triangle.***

end of the log as shown in illustration 8.8. Hold the square tightly against the chalk line and use the dividers to measure between the inner edge of the short leg of the square and the trim line on the exterior face of the log. Transfer this distance to the other face by holding one point of the fixed dividers against the blade of the square where it projects beyond the log end, letting the other point mark the wood. Draw the new trim line so that it runs through this mark; the new cut line should then be drawn exactly parallel.

Now turn the log onto its concave (bottom) edge, dog it in place as you did when hewing it, and with a flexible ruler extend the vertical trim and cut lines over the curved upper surface of the log as in photo 8.7. Then trim off both log ends with a saw (photo 8.8). I use a five-point crosscut. On the fresh ends, mark out new horizontal centerlines by connecting the ends of the centerlines chalked on both hewn faces (see photo 8.9), then draw vertical lines perpendicular to the horizontal centerlines you've just made, using a triangle as shown in photo 8.10, or the 90-degree edge of the template. These vertical lines will help you line up the slope of the notch across the log end. As may be seen in the photo, the vertical lines do not have to bisect the log.

Photo 8.11 *Using the template at the log end to draw the sloped upper notch.*

Photo 8.12 *Drawing the inner side of the upper notch.*

Photo 8.13 *Sawing the vertical cut for the upper notch.*

Leaving the log in place, position the template as shown in photo 8.11 so that its vertical edge is aligned with the vertical line drawn on one of the log ends, and the angled line slopes downward from the *interior* face of the log to the end point of the line drawn on the opposite (exterior) face. Mark a line along this edge to indicate the cutout depth of the upper notch. On the interior face, connect the end point of the sloping line—its highest point—with the cut line by using the 90-degree edge of the template, or a ruler as shown in photo 8.12, and draw a line perpendicular to the cut line joining the two points. Mark the lower notch the same way. The sloping line that extends between the cut line and the log end on the interior face should be parallel and level with the sloping lower line on the exterior face. Now repeat these steps at the other end of the log. The complete notch—both upper and lower cutouts—is now laid out and ready to cut.

CUTTING HALF-DOVETAIL NOTCHES

Begin by sawing down to the sloping face of the upper cutout, as shown in photo 8.13. Stand facing the exterior face of the log. Hold the saw at an angle and saw down along the log's cut line until the tip of the saw passes through both hewn faces of the log. Then go to the other side of the log and continue sawing from the interior face. Alternate sides until you reach the depth-of-cut lines for the upper notch on both faces. Make several more angled cuts parallel to this line, then remove the bulk of the waste wood with a heavy chisel and mallet. I use a 2-inch-wide chisel, held bevel down at first, and drive it in from the log end toward the scoring cuts. Don't try to chisel the waste down to the sloping line all at once. Stay about ⅛ to ¼ inch above the line. If the wood begins to split inward, below the plane of the slope, chisel across the cutout (see photo 8.14) or from the cut-line portion of the cutout toward the trim-line side. To pare the final surface, first use the chisel to chamfer (bevel) the edges of the wood down to the marked outlines of the cutout (photo 8.15), then remove the waste by chiseling across the grain (photo 8.16). Use a very sharp chisel, held bevel up. Check for high spots on the face by holding a straightedge across it, as shown in photo 8.17. Mark them with a pencil and continue to carefully pare wood away until the surface is flat. Repeat at the other end of the log.

Now for the lower cutout. Rotate the log so it rests on its convex (upper) edge. The notches you just cut should be on the bottom. Secure the log in place as before. Saw straight down along the cut line at one end of the log to the point where the sloping depth-of-cut lines intersect it. As with the earlier cutout, saw a few additional parallel cuts for waste removal (photo 8.18) and chisel out the unwanted wood (photo 8.19). You may want to use a hand adz for removing the waste wood from the bottom cutout. The adz should be light in weight, with a short handle, and have a flat blade with a rounded edge and the bevel facing up (toward you when in

(continued on page 142)

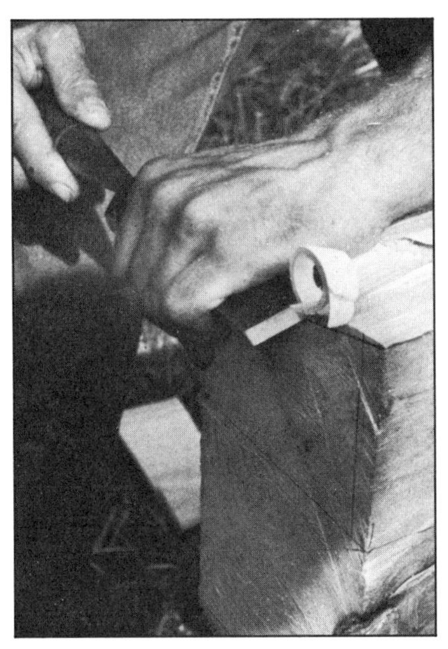

Photo 8.14 *Paring to the trim line.*

Photo 8.15 *Paring to the side outlines.*

Photo 8.16 *Chiseling across the notch.*

Photo 8.17 *Checking for high areas, using a straightedge.*

Photo 8.18 *Scoring the bottom notch.*

Photo 8.19 *Removing most of the waste with a chisel.*

Photo 8.20 *Shaping the bottom notch with an adz.*

Photo 8.21 ***Pocket notches for floor joists mortised into a sill log.***

use). First, chamfer the edges of the cutout with a chisel as before, then draw a series of freehand pencil lines across the chamfer at the sides and ends of the notch to show where material will be removed. Pick up the adz, stand on the trestle so you are straddling the log, and begin hewing away the waste in thin shavings, using strong, controlled swings as in photo 8.20. It helps to brace your elbows tight against your knees or thighs to keep the adz from cutting too deep. When you have reached the depth-of-cut lines, hold the adz slightly askew and keep swinging in order to pare out a slight depression about 3 inches back from the log end. This slightly concave surface, actually no more than ⅛ inch deep, will make the notch easier to mate with the one on the log below. Cut out and trim to size the cutout on the other end of the log, and the entire notching procedure is complete.

Small errors, imperfect hewing and log warpage can mean having to make a few corrections. These adjustments are usually made on the log wall, after the notches are test-fitted, as described in chapter 9, Raising.

JOIST NOTCHES

Joist notches are made by mortising slots into the wall logs to accept tenons shaped on the ends of the joists themselves. Structurally, joists require sound materials, adequate beam cross section and careful notching. Joists often function as tie beams, restraining rafter thrust on gable roofs. These joists require locked notches; joists placed at the sill can simply be set in pockets open at the top. Ground-level joists are often left full size and set in shallow mortises, although a tenon may be used as a means of adjusting the height of the top of the joist. Joists notched with pocket notches (photo 8.21) don't require spikes or pegs, and, in fact, they should not be spiked if they are shallow, because they might split.

An upper-level joist can be tenoned on its lower or upper surface, or a double notch may be used. The tenon usually extends through a mortise in the wall log, then is spiked through its top surface, locking the joist to the wall log below. An alternative to spiking is to use dovetailed joists set into sloped wall mortises.

Joist tenons are laid out and cut much like corner notches. A major consideration is that centerlines for layout be set at a standard distance from the supporting members beneath them, so the joists will all be the same height on the top surface. The depth of single notches should not exceed one-third the joist thickness, and total notch depth (where there are both top and bottom notches) should not exceed half the joist thickness.

The upper sections of notched joist beams often cross the top of the wall logs at a greater height than the chinking dimensions. Standard chinking can be maintained by taking a shallow notch off the top of each joist beam, where the chinking crosses, or by cutting through-mortises across the bottom of the succeeding wall log.

Joist tenons made from green timber shrink in width during seasoning. This can result in a gap of as much as ⅛ inch on each side of a 6-inch-wide joist. Mortise widths are stable, but crossgrain shrinkage may open further the gap between vertical cuts on tenons and the mortised log. The shrinkage effect for ground-floor pocket mortises is negligible; however, gaps in second-floor and attic joist notches cause air leaks, which must be carefully caulked. Gaps can be minimized by using seasoned logs for the joists.

Pairs of joist mortises must be laid out between opposite walls with precision so that the tops of the floor joists are all level with each other. Be sure also that the joists spanning the ceiling do not interfere with the headroom of people standing in the living area below. When laying out mortises, use a standard depth measurement taken from the snapped centerline on the side of the log. This will ensure level joists, regardless of variations in the thickness of joist logs. If you have construction delays, protect joist and corner notches from rain with scrap roofing metal or other materials.

Mortises can be made with the joist wall logs in place on the building, but in the long run, it's quicker to remove the log after the corner notches are properly fitted and level centerlines are established, and mortise the log on the ground.

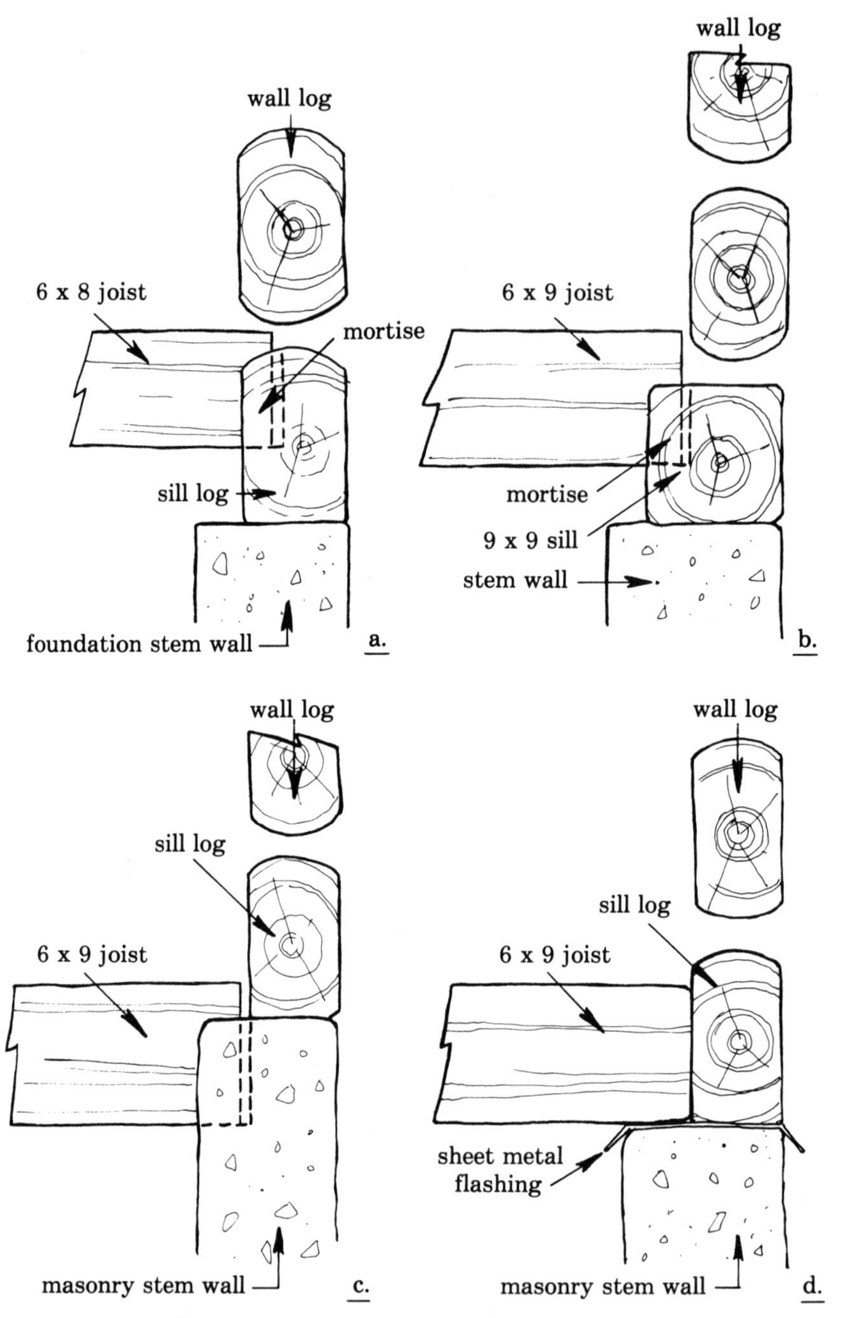

Illus. 8.9. *Joist notches (ground floor). (a) Ground-floor joist mortised into standard-width wall log. Note ½-inch air space behind joist. (b) Ground-floor joist mortised in 9 × 9 sill log. One advantage is that joist height is not limited by wall log dimensions. (c) Joist set in pocket cast into masonry stem wall. Advantage is lowered floor level. (d) Joist set on foundation stem wall. Sheet metal flashing between masonry stem wall and logs.*

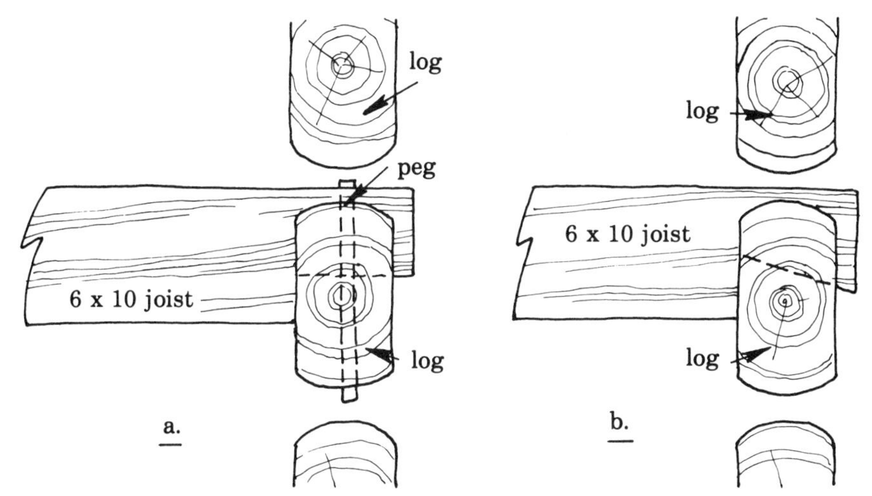

Illus. 8.10. ***Joist notches (upper level). (a) Upstairs joist notched into wall log and secured with 1-inch-diameter wood peg. A countersunk 8-inch-long spike can also be used. (b) Upstairs joist dovetail-notched into wall log.***

SPLICES

Splices are notches used to connect short lengths of timber. Splices joining sections of wall logs are generally cut vertically, while joist splices must be horizontal and centered above posts or girders. Splices can be joined with spikes, pegs or bolts. I use bolts with large washers because the nuts can be tightened as logs season and shrink. Splicing can be a very effective method of utilizing odd log lengths, but because spliced pieces are never as strong as whole logs, several rules should be observed: Never use more than one splice for one complete corner-to-corner log. Avoid using splices for consecutively stacked logs on the same wall. Logs with joist mortises or door and window lintel or sill notches should not be spliced. Corner notches, blocking and door and window frames can help to support the load of spliced logs.

The most basic splicing joint is the simple lap. A locked splice is more involved, but takes considerable load off the bolts or pegs which fasten the joint. A splined joint is a little easier to fashion, but still provides good holding power (see illus. 8.11).

Wall splices should overlap at least 2 feet, but for joist splices above a girder, the overlap can be 6 to 12 inches. The cutting procedure is similar to that for corner notches. Saw along the vertical cut line to the marking lines, then make parallel scoring cuts every 3 to 5 inches. Remove the bulk waste with a 2-inch chisel and a mallet. Chisel a chamfer to the

(continued on page 148)

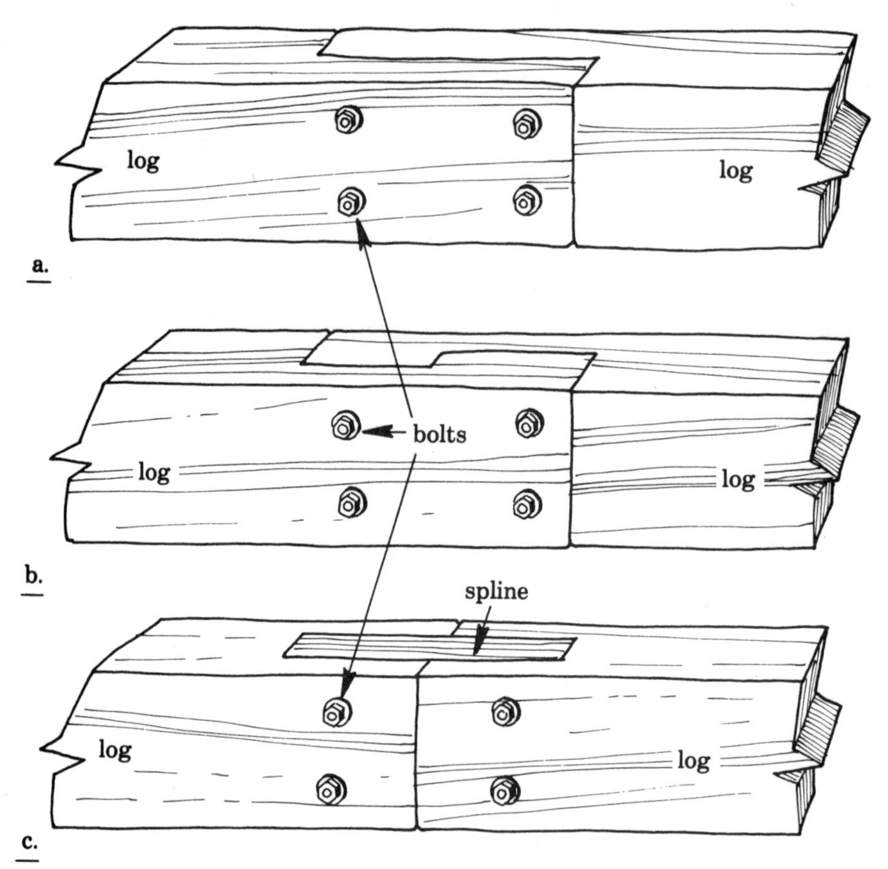

Illus. 8.11. *Splice joints. (a) Lap splice. (b) Lock splice. (c) Spline splice.*

Photo 8.22 ***Floor joists reinforced by a central girder and supported at the midpoint by a post set on a concrete pad. Aluminum flashing acts as a moisture barrier and termite shield.***

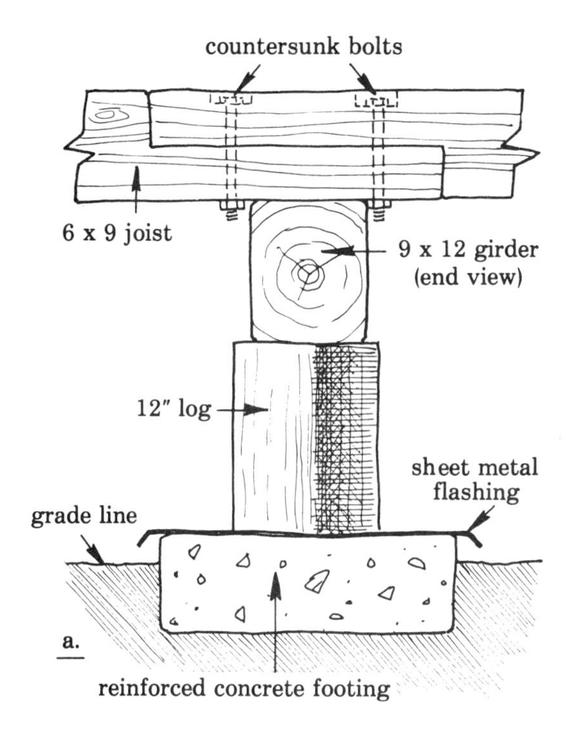

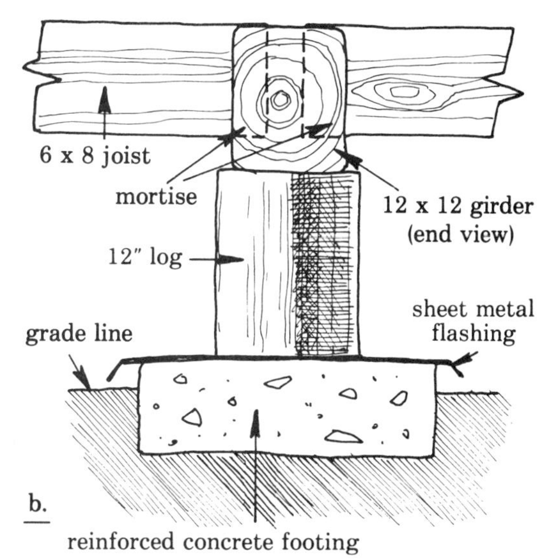

Illus. 8.12. *Spliced bearing timbers. (a) Spliced ground floor joist supported by girder. Pier can be log, concrete or masonry. Footing for log pier should be above grade, with sheet metal flashing between log and footing. (b) Large girder mortised to accept joists.*

outlines. Adz the surface flat or slightly concave. Work with or across the grain. Attach spikes, pegs or bolts after the spliced logs are fitted together on the log wall. Wall splices should be sandwiched between blocks so that the weight of the spliced logs does not bear on the spikes or bolt fasteners.

If you cut into joists or girders, remember that a beam's bearing capacity is reduced to that of the uncut cross section. Vertical posts under a notched beam can help take up the load. See chapter 12, Finish Work, for a description of lintel notches and chapter 10, Roofing, for plate and rafter notches.

CHAPTER IX

RAISING

One of the most exciting parts of logbuilding is raising the walls. By the simple, repeated process of laying one hewn and notched log across two others, a rough pile of timber is transformed into a unified structure, a home. You experience the reality of the interior space for the first time and begin to see how the new building will relate to its surroundings.

To get the logs in place, you must rely on equipment such as rope, pulleys, inclined planes and levers. The raising process includes several steps in addition to moving logs. Adjustments in the depth of notches is inevitable, and you may want to treat notch faces with a preservative before they are joined. Sills should have good metal flashing, which can be tacked to the bottom of the sill log before it's set in place. As the logs go up, check the structure to ensure that the walls are plumb and square and the notches are level.

Take your time. Think through the procedures you'll follow, and work within comfortable, safe limits. Proper tools should be on hand.

Photo 9.1 ***Logs becoming a house.***

Photo 9.2 *Homemade timber cart. Secure the log with a chain and a light-duty load binder.*

Photo 9.3 *Hauling a log by hand.*

Photo 9.4 *Temporary scaffolding.*

Pathways, scaffolds, and work areas must be clear of debris and dry to assure good footing. Don't attempt to lift or carry heavy logs when you're tired or feeling rushed. Whenever there is a question of safety, stop and think the situation out. Get help or needed equipment. Don't push your physical limits while moving logs; the consequences are too serious.

Moving logs and working on unfloored joists is dangerous. Before you begin raising logs for the walls, put in temporary flooring—as soon as the joists are in place. Use sound, cheap 1-inch planking. I start flooring about 6 inches from walls and leave a 1-inch gap between boards to allow the joists to dry better after a rain. For a large structure you may prefer to use 3- or 4-foot-wide catwalks around the inside perimeter of the walls, leaving the interior open. Do not install permanent subflooring

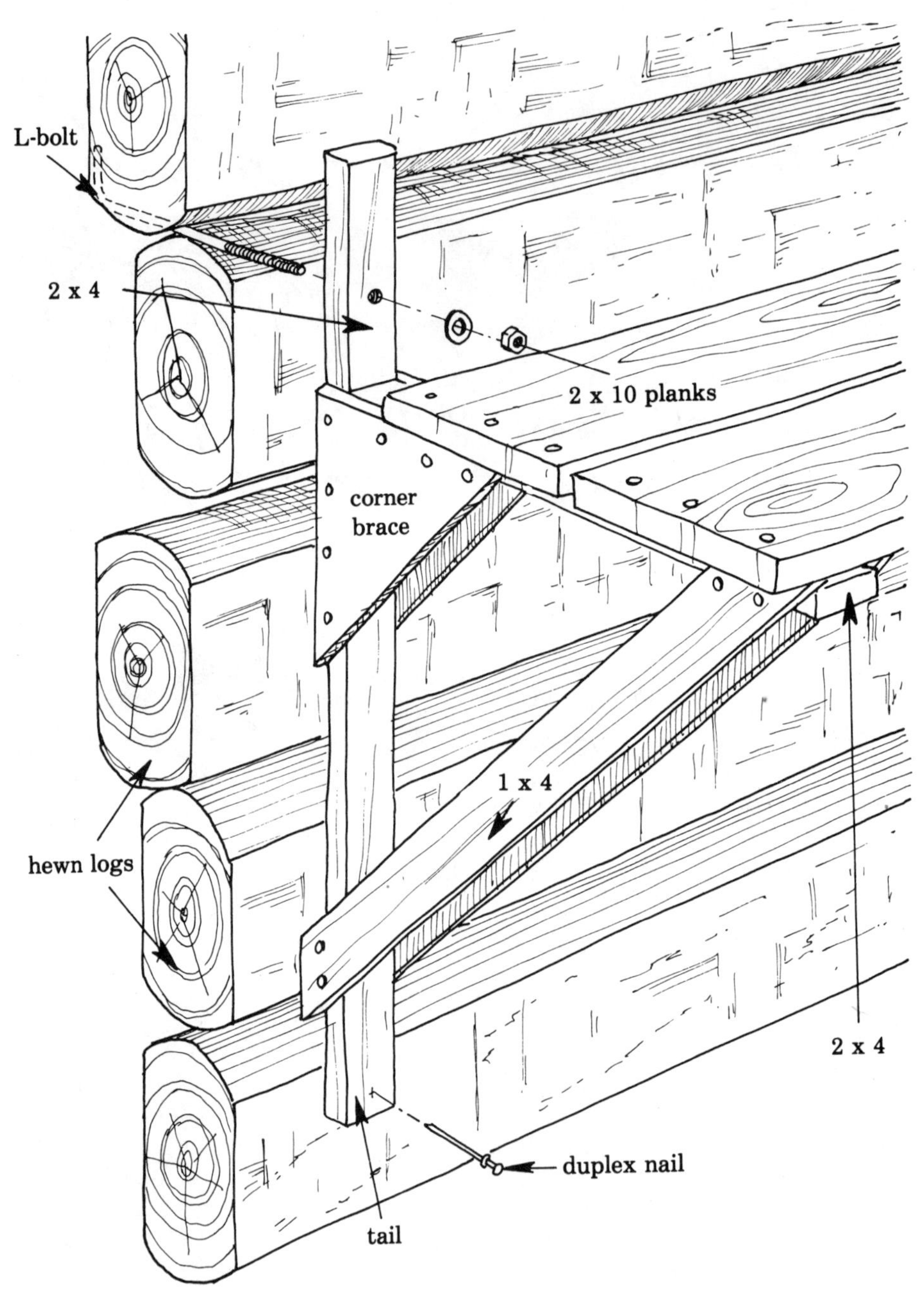

Illus. 9.1. *Hanging scaffold. The L-bolt fits through the chinking space between the logs. Use double-headed (duplex) nails to anchor tail of scaffold to logs.*

at this time, because without a roof, rainwater gets trapped between joists and flooring.

Scaffolds are very useful after walls reach chest height (see photo 9.4). They make it possible to stand near the skid pole wall while hauling raising ropes without running the lines over the wall rim, and they make it easier to move logs onto the cross walls. Scaffolds make it easier to make notch corrections.

RAISING LOGS

The time-honored method of raising logs is with skids, which are simply poles propped against the log wall. The logs are pulled up these inclined poles with ropes. I've used 5-inch poplar saplings, but any smooth, straight poles in sound condition are suitable. Skid poles should always be debarked and greased with lard or tallow (which won't stain the logs) to reduce friction. Flatten the upper end of each pole with an axe, so it can be securely nailed to the top of the log wall; otherwise the poles could roll sideways during lifting. As a precaution, I also nail short braces from both sides of each skid pole to the top log. The skids must not protrude more than a few inches above the wall; if necessary, bevel the tops to reduce the height.

The length of the poles should be adjusted so that the angle of the inclined plane will be no more than 45 degrees when the highest logs (the top plates) are being raised. On a sloped site, adjust the lengths so the poles lay roughly parallel to each other. If possible, use two sets of skid poles, one for the long sides of the building, the other for the short sides. It isn't hard to push a log from the near wall straight across to the wall opposite (if the log extends well out beyond the two supporting walls), but turning corners with a log on the walls is tricky.

Center the log being moved when it is at the base of the skid poles. If the raising height is less than 5 feet, you will probably be able to push the logs up the poles while standing on the ground, but always stand to one side while pushing on the log end; never stand beneath the log. Use two ropes, one fastened at each end of the log, for better control of the lift.

While it's possible to use skid poles single-handed, two people make the work much easier, and four people make an ideal team for raising logs over 16 feet long. Small crews are more efficient than large groups. Although things move quickly with lots of help, risks also begin to increase. It's hard to control the action with too many workers, and there's a greater chance that someone will get hurt.

Use block-and-tackle rigs or a "come-along" (a ratchet and lever device for tightening cable) to help pull the logs up the skid poles. I use come-alongs because the ratchet mechanism acts as a safety device, but the

Photo 9.5 *Using skid poles.*

limited cable lengths require the use of holding ropes partway up the skids while the come-alongs are reset. You can tie the come-along to the logs on the wall opposite the skids, then pass the cable (or rope extension) over the skid wall, pulling from inside the house. The problem with this arrangement is the friction caused by ropes passing over the rim of the wall. Reduce the friction by tacking greased flashing metal to the top of the wall, where the rope runs. This method is slow, but sure. A tractor or motor winch located at the side of the house opposite the skids (you'll need long ropes) will greatly speed up this procedure.

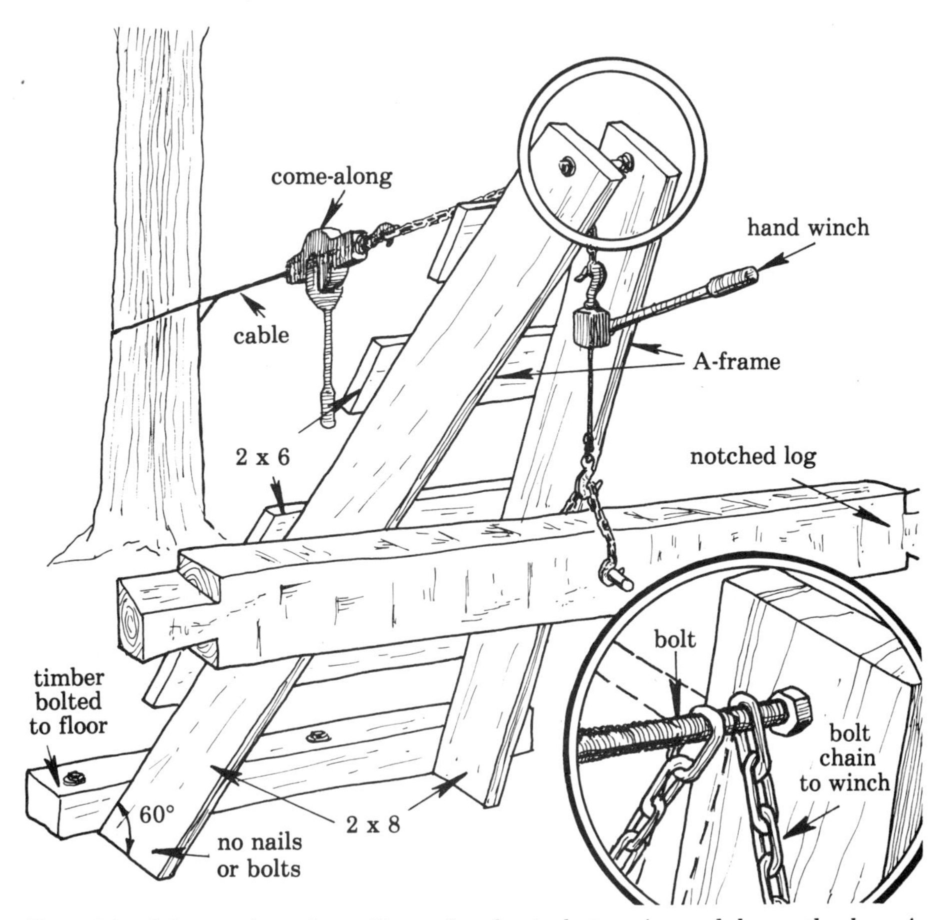

Illus. 9.2. *A-frame shear legs. Use a hand winch to raise and lower the log. A "come-along" ratchet hoist adjusts the angle of the frame (60 degrees is optimum angle for lifting). Secure the cable end to a wall log, tree, or other substantial anchor. The base of the A-frame rests against a timber bolted to the floor or anchored to the ground. The A-frame base requires no fasteners.*

Other raising methods are needed if the site is not suitable for skid poles, or if the logs are too heavy to be handled with the personnel and equipment on hand. I have always used skid poles, but other logbuilders use homemade cranes, cable strung between trees or poles, or a trolley running on a rail that crosses the top of a scaffold. Power cranes and forklifts can also be used, but they damage the site.

Once the log is pulled to the top of the skid poles, lift it slightly so that it rests on the adjacent cross walls. Use poles or peavies as levers if the log is heavy. When the log is directly above the log on which it will rest, rotate it so its notch is seated on the log below. Setting a log in place is a tense moment, as you anticipate the heavy clunk and hope for perfectly fitted notches. When moving the log into place, it's best to sit straddling the adjacent walls if you don't have scaffolding. Be sure to hold the log in such a way that your fingers or toes won't be crushed by a chance tight fit between logs. Sometimes a few blows with a heavy sledgehammer are required to persuade a log into its final position. Joist logs, in particular, must be fitted tightly.

Always inspect both notches for a proper fit. Try to wobble the log by grasping the top and shoving it back and forth. Check the inside edges of the notch. The log should sit solidly without spaces at the faces of the notches.

NOTCH ADJUSTMENTS

Small errors, imperfect hewing and log warpage often mean you'll have to make a few corrections in the notch you first cut when the log was on the ground. These final adjustments are usually made with the log in place on the wall, but rotated upside down and secured with braces nailed to the supporting logs. Notch corrections on one end of a log will also slightly affect the level of the other notch. Height measurements, checked from accurate benchmarks, should be made every few courses so that the logs can be kept level.

The final notch depth should equal the gap between the two fitted logs, as measured with a scribe while the top log is in the notch-down position. If an edge or corner of the notch is too high, the entire notch must be lowered to that level. Mark the parts to be cut out, rotate the log so the notch is up, and use a straightedge placed across the face of the notch to be sure it remains even, with no high points. Trim the excess wood down to the scribed line with an adz or a chisel, and set the log back in place. It should fit perfectly.

Notches that are too deep can be shimmed at both ends or made to fit by hewing or sawing the contact area to a narrower thickness. The notches of succeeding adjacent cross logs can also be cut shallower than normal, as the total gain results in a correct fit. Notches will tighten up a little from the weight of the roof and succeeding courses of logs, so minute gaps are acceptable. But if logs warp during seasoning, a gap may open that will

Photo 9.6 *Setting a log in place.*

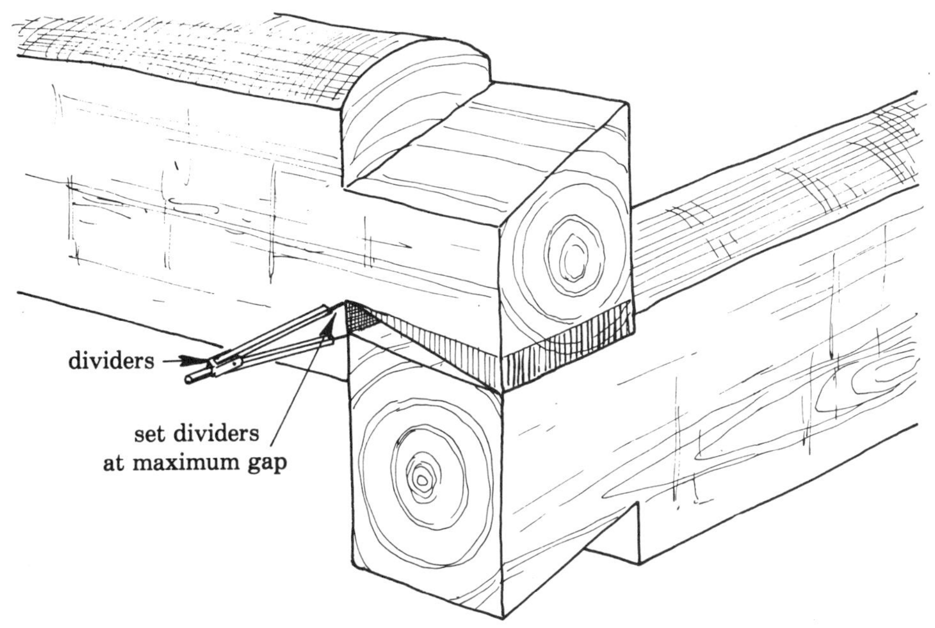

Illus. 9.3. *Scribe-fitting a poorly cut notch (extension not shown). Set dividers at maximum gap, then scribe around notch following top edge of lower log.*

Photo 9.7 *Temporary braces used at door openings also support a lintel log.*

Photo 9.8 *Pieced-together logs braced in place.*

have to be filled with caulking. Once a log is properly fitted, nail a short piece of wood to the preceding log to hold the notch in place until the next cross log locks it in position.

Photo 9.9 ***Spliced logs are first blocked and braced, then they are spiked, doweled or bolted together.***

PIECED AND SPLICED LOGS

If you can't find a long enough log, you can piece two short lengths together by cutting a lap joint in each and spiking the two pieces together. For hardwood, drill out a hole for the spike, but for pine, just drive the spike into place. Thorough instructions for splicing log sections are given in the chapter on notching. These pieced-together sections are easy to handle and produce a durable log, but require extra attention during assembly. It's best to put one piece in place on the log wall and secure it temporarily to the log below with braces before adding the second section.

In addition, window and door openings always result in wall sections of short, unsupported logs, whether you build up the logs around the openings or cut out the openings after logs are in place. (If you cut out the openings after the logs are in place, do it after the walls and rafters are complete, and never cut the sill or top plate.) Support the short pieces carefully, until window and door frames are complete and tied in, by running vertical braces along the unsupported ends of the logs and tying each of these braces to the floor with a diagonal brace to keep the work steady. When you are building up short pieces around an opening, make sure the logs don't drift out of plumb. To prevent this, snap a chalk line down the center of each unsupported log end and check every few courses with a level held up against the line. Short pieces between two adjacent openings are very unstable because there are no notches to secure them. In fact, I withhold these logs until I begin work on the door and window frames.

It's important that when you look at a finished wall, each horizontal log appears to be cut from a single piece—even if your pieced-together sections are made from many different logs. Maintain this effect when you fit the sections in place by assuring that all the short pieces are the same size as others on their level.

CHAPTER X

ROOFING

After the foundation is laid and the walls are notched and raised into place, much of the heavy work is done, but until the roof is framed and covered, all the logs and notches are open to the weather. When the roof is finished the place really begins to look like home.

Various materials can be used to frame the rafters for a log house: hewn timbers, standard-dimension lumber, pole rafters or purlins. Dimension lumber works fine, but the other methods are more in keeping with the rustic quality of a log house, and much cheaper than using commercial material. Pole rafters, which are thinner than wall logs, are positioned much like standard rafters. On a gable roof they are notched into the plate logs and sloped from opposite walls so they meet at the ridge line, then they are sheathed and covered with roofing material or overlaid with horizontal lath and shingles. Purlins are horizontal supporting beams laid between opposite gable ends of the structure. Purlins transfer the load of the house to the gable ends, eliminating outward thrust on the walls that is an inherent factor with rafters.

Photo 10.1 ***Roof detail at the Gott homestead.***

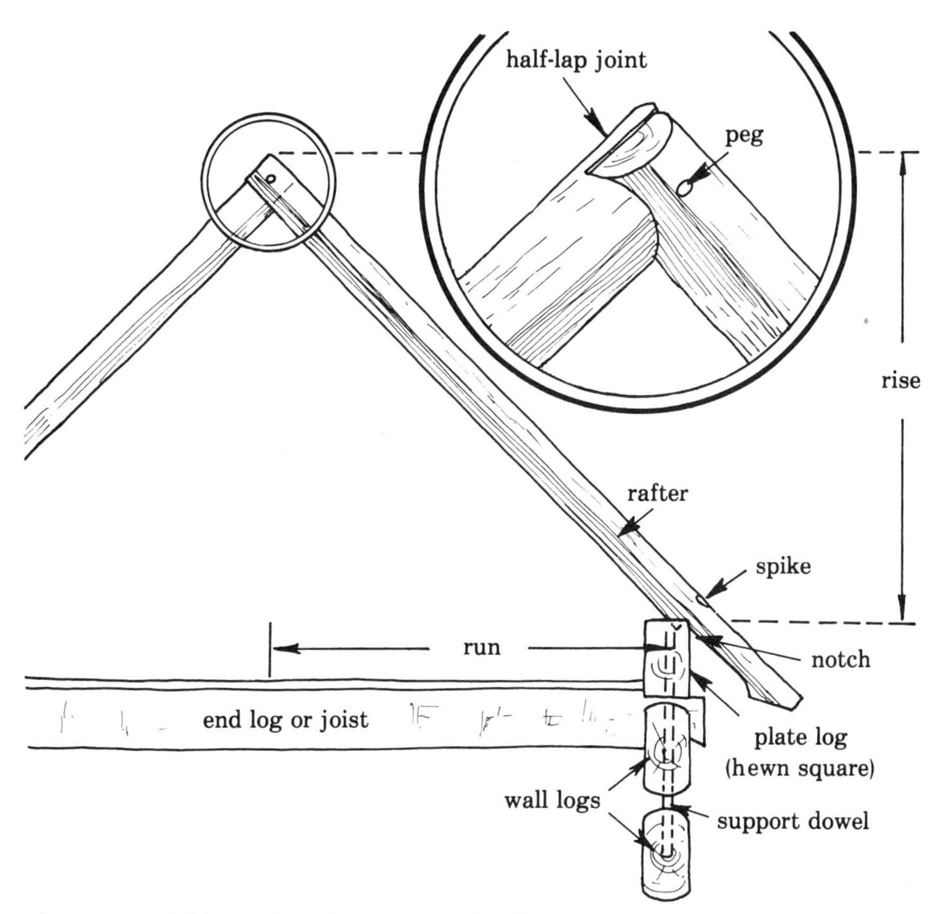

Illus. 10.1. ***Gable roof. Rafter tops are half-lapped and pegged together in pairs. Bird's-mouth notches in the rafters are spiked to the plate logs.***

POLE RAFTER FRAMING

Despite the advantage of purlins, we decided to use rafters on our roof. We used poplar saplings for our rafters and connected each pair at the peak with matched half-lap notches pegged with a wooden dowel. There is no ridge pole. At the plate (the top wall log) we secured each rafter with a load-bearing notch and a 6-inch spike. Rafter thrust against the plates is contained by tying the plate logs into the attic joists. These joists span the two long walls, plate to plate, acting as collar beams to prevent the walls from spreading. The attic gable wall also helps to hold the roof together. The rafters are held in place by the horizontal shingle laths and by four diagonal braces that connect the midpoint of the plates to the peak of each gable. Since our attic is a storage area, we insulated between the upstairs ceiling and the attic flooring and left the attic unheated.

Nearly any tree species that has straight saplings with very little taper can be used for rafter poles. Try to gather a set of poles that vary little from one to another. Often the poles seem much straighter in the woods than they turn out to be when you get them up on the hewing trestle. Most have at least a slight crown. In extreme cases, you may have to do a little hewing to straighten them up.

Our rafters, 14 feet from rafter tails to roof peak, taper from between 5 and 6½ inches in diameter at the tails to 4 or 5 inches at the peaks. We left our rafter poles round, but you may prefer to hew one face to provide a uniform surface for the lath and shingles. If you sheathe the rafters for a standard roofing system (plywood, building paper and asphalt shingles) you must hew the top surface of the poles to assure a smooth nailing surface.

LAYING OUT AND CUTTING THE RAFTERS

To accurately mark and cut notches for the rafter poles, you'll need to make a pattern. Because rafter poles are round in cross section and invariably somewhat bowed, they present a challenge to the builder, but by using a carefully made pattern as a guide, the poles will fit snug on the plate notches, and their top sides will all be the same height above the plate notches. Even though the poles may bow somewhat, the variation will only be an inch or two, very minor considering their length of 14 or more feet.

The pattern we used was cut in a 1 × 6 board, 14 feet long (the length of our rafters, including overhang). Make the pattern the same length as your rafters. At one end, make a diagonal cut to match the angle of the roof peak; at the other end, cut out a section to match the notch face in the plate log.

The simplest way to reproduce the angle of the peak is with a framing square. Set the square against the 1 × 6, so that the board crosses both legs of the square, and move the square until the board crosses the short leg at the rise and the long leg at the run (see illus. 10.1). If the roof pitch is 9 on 12, for example, the board should cross the long leg at the 12-inch mark and the short leg at the 9-inch mark. Measure the distance between the two points where the board crosses the legs. Mark the midpoint of this distance and draw a line on the pattern board from the mark to the inside corner of the framing square where the two legs meet. This line corresponds to the angle of the roof peak.

The pattern for the notch is set up so that even using rafter poles of varying diameters the finished rafters cut on the pattern will all rest tightly in the plate notch, with their upper surfaces level. The key is to cut the rafter first to a set thickness above the notch, then, if the rafter is large, to hew it down along a secondary curve on the pattern until the notch face on the pole is the same length as the notch face on the plate notch. If the rafter

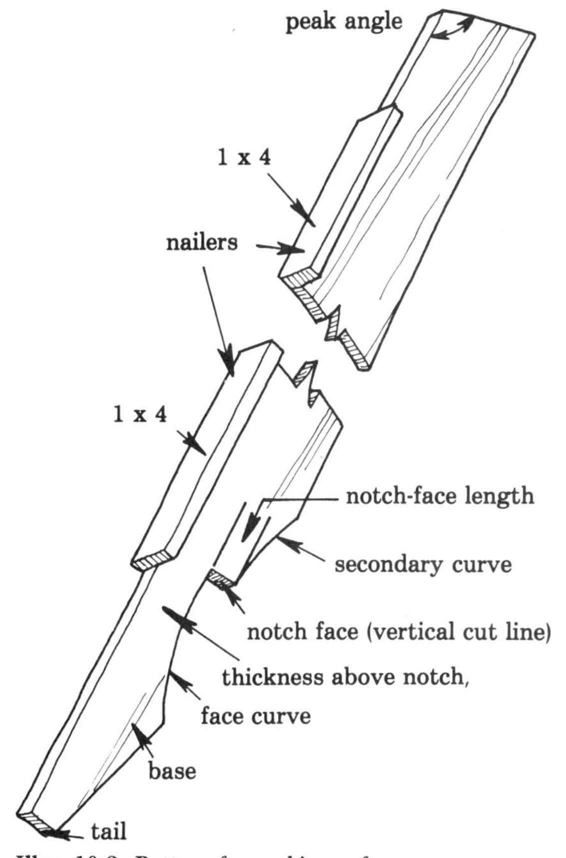

Illus. 10.2. *Pattern for making rafters.*

is thin, the rafter notch face will be less than normal, but because you first cut the rafter to a set thickness the top of the rafter will still be level with the others when it is mounted on the plate log.

First make a vertical cut through the edge of the pattern board, toward the edge where the nailers will go (see illus. 10.2). Stop when the uncut portion of the pattern equals the predetermined thickness of the poles above the notch. Mark a sloping curve on the face of the pattern board from the same edge you've just cut to the bottom of the notch face (the vertical cut line), and cut along the line with a chisel and mallet, or an adz. This sloping cut creates the face curve. The exact curve isn't critical, as long as there is room enough for the face curve to clear the plate when the rafters are mounted.

Once you've made the notch cut and the face curve, mark the depth of the plate notch on the template along the vertical cut line. This is a measurement you develop somewhat arbitrarily, taking into consideration

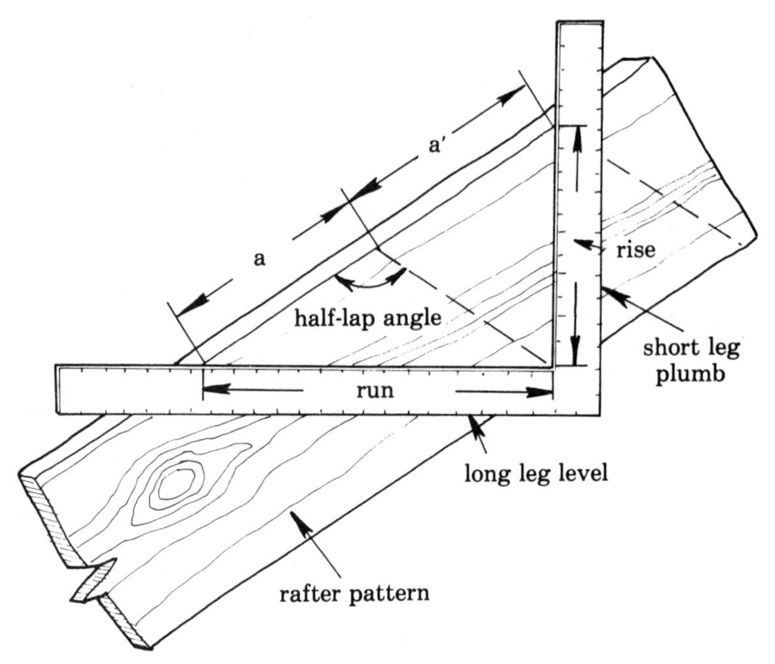

Illus. 10.3. ***Determining the roof peak half-lap angle. Position the framing square to equal the angle of the roof pitch. Divide the hypotenuse of the triangle formed into two equal parts. The broken line in the diagram intersecting the hypotenuse and the 90-degree corner of the framing square is the correct angle of each half-lap.***

the dimensions of the wood you're working with. You need to have adequate bearing surfaces but you must avoid cutting too deeply into either the plates or the rafters. A notch depth of 1½ inches usually works fine. Cut a secondary curve to that mark if you have to, for instance, if the template is wider than the rafters you will use, cutting from the board edge to the notch face as shown in the illustration.

The final base line is not essential to cutting the rafters; it's for looks. Mark the base on the pattern board by first making a mark at the end of the board that divides the width of the board in half, then cutting a straight line from the beginning of the face curve (where it meets the board edge) to the line you just marked at the end of the board. On the rafter, this cut end is called the rafter tail.

Complete the pattern by adding two 1 × 4 nailers, each 2 feet long. Place each nailer several inches from opposite ends of the pattern board (along the uncut edge) and nail through the 1 × 4 face into the edge of the pattern board. One edge of the nailers should be flush with the face of the pattern board. The other edge should protrude beyond the pattern board face to form a lip.

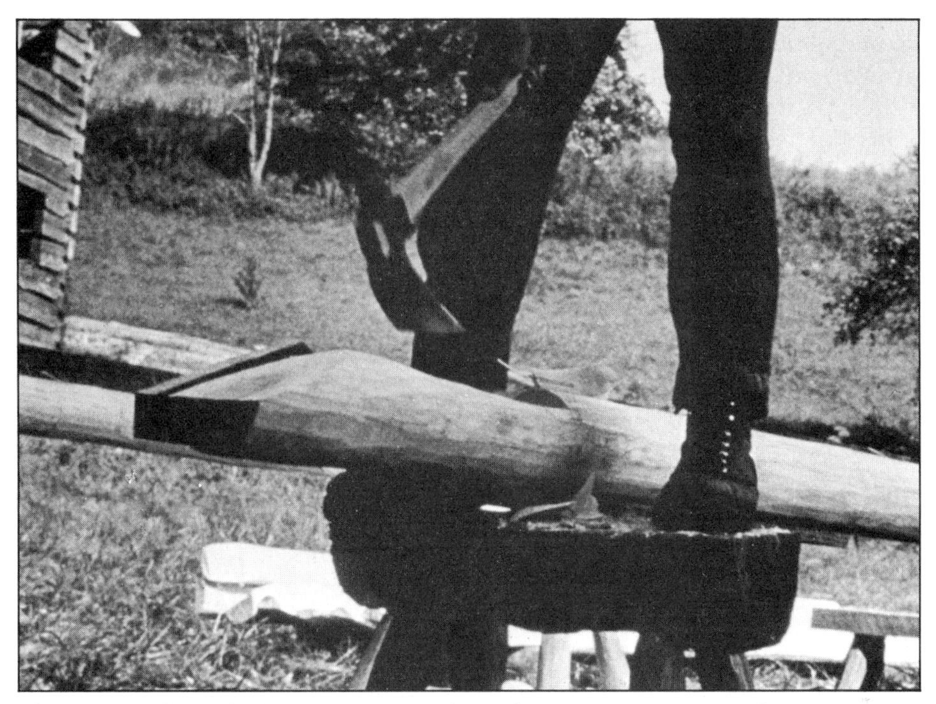

Photo 10.2 *Adzing the face curve on a rafter pole.*

Photo 10.3 *A finished rafter tail shown upside down.*

Set the face of the 1 × 6 on hewing trestles, with the lip formed by the nailers extending up. Place a debarked rafter pole on the 1 × 6 with the crown against the nailers, and nail through the face of each nailer into the rafter. On the roof, the crown of the rafter will be up, supporting the lath. If you plan to hew one side of the rafters, do so before putting the rafter in the pattern. You can also hew on all four sides, but then you'll need considerably larger poles.

First cut the half-lap for the rafter peak. Cut off the thin end of the pole, following the angle of the pattern board end. Measure in from the end a distance equal to the diameter of the other rafter of the pair, and make a second cut parallel to the first, but only halfway through the rafter. Make several similar cuts between the two saw cuts (to ease waste removal), and chisel out the unwanted wood, starting the chisel cuts from the rafter end.

To make the notch cuts, first turn the pattern/rafter assembly a quarter turn so it rests on the nailers, and the edge of the pattern board is up. Make a vertical cut following the pattern, stopping where the vertical cut of the pattern meets the curved surface of the face curve. No matter what size pole you cut in this manner, the distance between the depth of the notch and the nailers will remain constant. Now cut out the face curve with an adz. The face curve of the pattern board serves as a guide, assuring that the curve will end exactly at the bottom of the notch face. If necessary, use the adz also to hew out the secondary curve. Finish off by sawing the base line, then check that the rafter thickness above the notch face, the notch-face length, and the overall rafter length are correct.

If the rafter diameter is greater than the width of the plate notches, the notch sides will also require some hewing (since these are done before the logs are set in place, they are cut to a single, standard dimension). Use an axe or an adz. Hew both sides of the rafter pole, so it will be centered over the plate notch.

PLATE NOTCHES

The plate notches (see illus. 10.4) carry the main thrust of the rafters. The rafter notch face rests on the plate notch face, and part of the rafter's face curve rests on the plate's beveled outer edge. (The secondary rafter curve, by the way, must provide enough clearance to let the face curve sit on the beveled edge.) Both the inner face of the plate notch and the beveled outer edge of the plate match the roof pitch.

Plate notches should be cut out before the plates are set permanently on the walls. Careful location of notches is a must since the distances between the notches on one plate and those on the plate opposite define not only the span of the roof but the pitch as well. To locate plate notches properly, first determine the roof span by measuring between the two log walls upon which the plates will rest and then add some arbitrary measure-

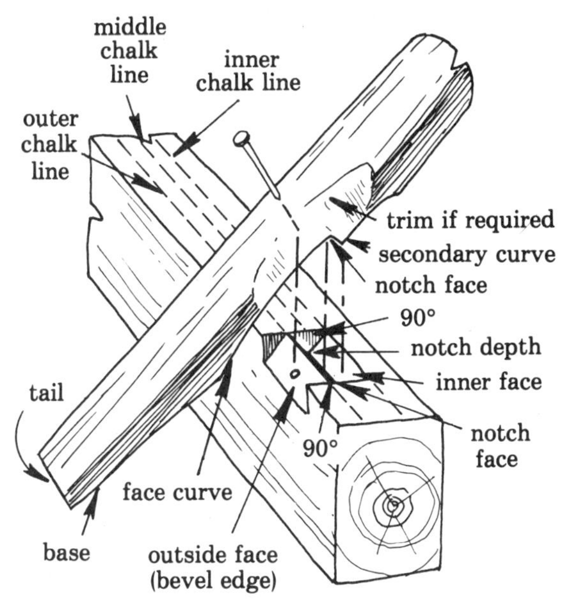

Illus. 10.4. *Plate notch layout.*

ment equal to a portion of each plate's width, measured from the top edge of the plate's interior face toward the exterior. On each plate, this distance away from the interior edge determines the locations of the plate notches, and thus marks the spots where the rafters make contact with the walls. When choosing this distance remember also that the span you are creating determines in turn the roof pitch that will result, so work carefully with your building plans before you actually start laying down lines on wood.

Since plate logs sometimes warp before they can be installed, the most accurate way to mark the notch locations is with a snapped chalk line. At both ends of the plate, measure from the interior face toward the exterior the distance you have chosen for the notch locations, and then carefully snap a straight line running between the points, along the length of each plate. This line is indicated on illustration 10.4 as the middle chalk line and is used as the basis of all subsequent notch measurements since the distance between this line and its mate on the other plate remains constant regardless of any bow or twist in either log.

The middle chalk line also indicates the location of the deepest part of each notch. The other two chalked lines shown in illustration 10.4—the inner chalk line and the outer chalk line—indicate respectively where the inner notch face meets the top surface of the plate, and how deeply the exterior face of the plate is beveled (at the same angle as the roof pitch) toward the center. The notch face slopes away from the beveled exterior face at 90 degrees to it (like a gable roof.) The inner face tilts up, parallel to the beveled face (also at the same angle as the roof pitch), and at 90 degrees to the notch face. When the notch is actually cut out, the area between the inner and outer lines, which includes the middle line, is removed.

To locate the inner and outer chalk lines on the plate, take measurements as follows: Place a ruler on a framing square so that the ruler crosses the two legs of the square at the same angle as the roof pitch. The ruler should cross the short leg of the square at the rise, and the long leg at the run. Maintain the same ruler angle on the square, but move the ruler until the distance measured on the short leg of the square equals the length of the notch face. You can now take direct measurements off this setup to find the depth of the notch. A right angle from the ruler to the corner of the framing square divides the ruler in two—equal to the distances from the middle chalk line to the inner and outer chalk lines. Measure these two lengths and use them to locate the other two lines. The shorter distance is from the notch bottom to the apex formed by the notch face and the beveled edge (middle and outer lines). The longer distance is from the notch bottom to the top of the inner face (middle and inner lines). Mark the position of the inner and outer lines at both ends of the plate, then snap lines between them along its full length.

Now mark the sides of the notches with a pencil and a framing square, by squaring lines across the top of the plate logs, using the middle chalk line as a guide. The widths of the plate notches should all be the same, decided in advance according to the widths of the rafter notches. Locate each notch at an appropriate spacing along the plate (say, on 24-inch centers), then lay out the lines for the sides on either side of each mark.

After all the notch widths have been laid out, hold a hand saw at the same angle as the roof pitch and cut in along one width line to the outer chalk line. Do the same along the second width line, then chisel out the beveled face. Use a sliding T-bevel or make a template to check that the beveled face slopes away from the vertical exterior side of the plate log at exactly the same angle as the roof pitch. Chisel out the notch face next, at 90 degrees to the beveled edge. Its length and the location of its deepest point are indicated by the outer and middle chalk lines. Then chisel out the inner notch face at 90 degrees to the face you've just cut. The slope of the inner face must also be the same as the roof pitch. As you chisel, frequently check your work by inserting a framing square, corner first, into the notch to be sure the lengths, depth, and angle are all correct. Cut out all the notches on one plate, then carefully recheck all your previous notch calculations before cutting into the second plate.

The notched plate logs can be secured to the gable walls with locking notches, steel rods or 1-inch dowels driven through several courses of logs. We also tied the plates to the attic joists with heavy steel brackets and lag bolts as shown in illustration 10.5. Thrust on the plate can also be restrained with collar ties and trusses.

Photo 10.4 *Plate notches for rafters, treated with water-repellent wood preservative.*

ERECTING THE RAFTERS

Two or three people can assemble and erect the rafters, working on temporary catwalks to position pairs of rafters over the corresponding plate notches, then crossing the peak half-laps. One person should hold the lapped pole ends together while another bores a 5/8-inch hole through them with a hand brace. Then, drive a dry hardwood dowel through the hole. Make sure that the rafter notches are positioned directly above the plate notches, and secure each pair of rafters with a temporary collar tie made of scrap 1 × 4.

Erecting the first pair of rafters requires three people and plenty of large hanging scaffolds. Prenail one end of a temporary brace to the 1 × 4 collar tie between the rafter pair. Two people on hanging scaffolds on the long walls hold the rafter notches in place while a third person lifts the rafter pair by walking a plank laid between scaffolds at the two gable walls. As soon as the rafters appear to be vertical, the two people out on the plate logs sink 6-inch spikes through the rafters and into the plate notches. (A 2-pound hammer works best.) Then the brace should be nailed to the gable.

Erecting the remaining pairs of rafters is easier. First, nail a temporary horizontal brace across the outer surfaces of the first pair of rafters. Then,

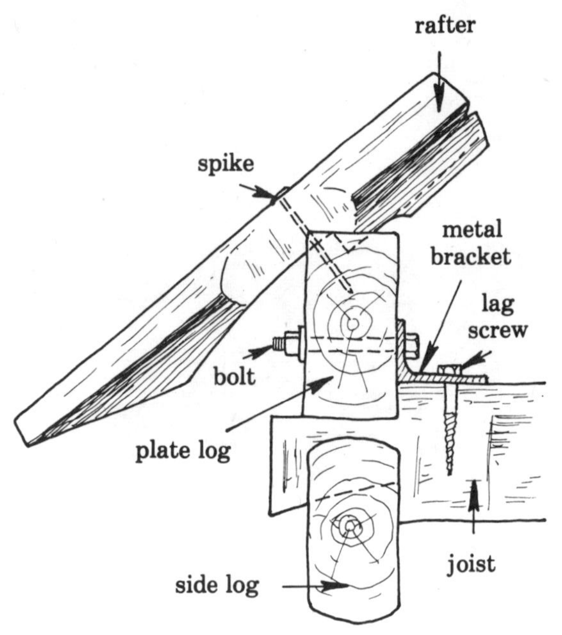

Illus. 10.5. *Plate notch and joist detail. Metal bracket secures plate log to joist or gable-end wall log.*

Photo 10.5 *The first pair of rafters braced in place.*

Photo 10.6 ***All the rafters in place. Furring strips are stacked in foreground.***

as you set succeeding pairs in place, nail them to the brace as well, in such a way that the spacing between pairs of rafters is the same at the peak as at the plate (where the spacing is determined by the notches). After a few pairs are in place and linked together you can rack them all plumb at once simply by removing the diagonal brace holding the first pair of rafters, and then plumbing any individual pair. Since all the pairs of rafters are linked together at equal distances—or are at least parallel to each other—when one pair is in position all the others will be in position as well. When the first group of rafters has been plumbed in this way, nail the diagonal brace back in place and proceed to erect another group of rafters following the same procedure. Continue until the job is completed.

PREPARING THE ROOF

Eaves gutters should be installed before shingles are applied. Attach gutters to a fascia board at the rafter tails, or hang them from the eaves furring strips with wire brackets. Use galvanized nails or brass screws. Homemade V-troughs and supports can be built of red cedar or other weather-resistant wood. Plain half-round galvanized gutters look okay, too.

Chimneys, flues and skylights should also be framed in before the shingles are applied. Be sure to flash and caulk carefully. Gable ends, which must include screened vents or windows, can be framed any time after the rafters are installed. Conventional 2 × 4 stud framing can be used to support lath for exterior gable-wall shingles, insulation and interior paneling. We used barn framing, nailed horizontally, to support the exterior sheathing of rough-milled board-and-batten siding, installed vertically. The gable ends of our annex, though, are timber-framed and filled with real plaster. The heavy mortised gable-frames of the annex support a hewn ridge beam that carries much of the roof load.

For a noninsulated roof, use 1 × 3 furring strips (lath) spaced so they will lie under the portion of each course of shingles exposed to the weather. The strips can extend as far as 1 foot beyond the end rafters, but nailing vibration becomes a problem at this distance unless a board is nailed to the ends of the furring strips. Such a piece, called a barge board, is usually a 1 × 4. It also protects lath ends and shingle edges from direct exposure to the weather. The laths at the eaves can be 1 × 6s. The bottom edge of the lowest one should extend about 1 inch below the rafter tails. The final pair of ridge laths are also 1 × 6s.

All the furring strips can be nailed at once, or you can put up one strip at a time as shingling progresses. Shingle roofs are steep and slippery, especially when the shingles have been treated with a greasy water repellent. To keep our footing, we shingled most of our 9-on-12 roof standing on scaffold planks between temporary rafter collar ties and applied the last shingles by straddling the roof ridge. You can also use a movable shingling seat as shown in photo 10.24 at the end of this chapter. A diagram of how to make one is shown in illustration 10.7.

SHINGLES

Shingle roofing may be the least practical element of traditional log buildings. Shingles are time-consuming to make and apply, and they can be a fire hazard, especially if moss starts to grow on them. It's impossible to know in advance exactly how long a shingled roof will last, and suitable material for the shingles may be hard to locate. Most of the old roofs in Europe and Japan that were originally shingled have been covered over with tiles.

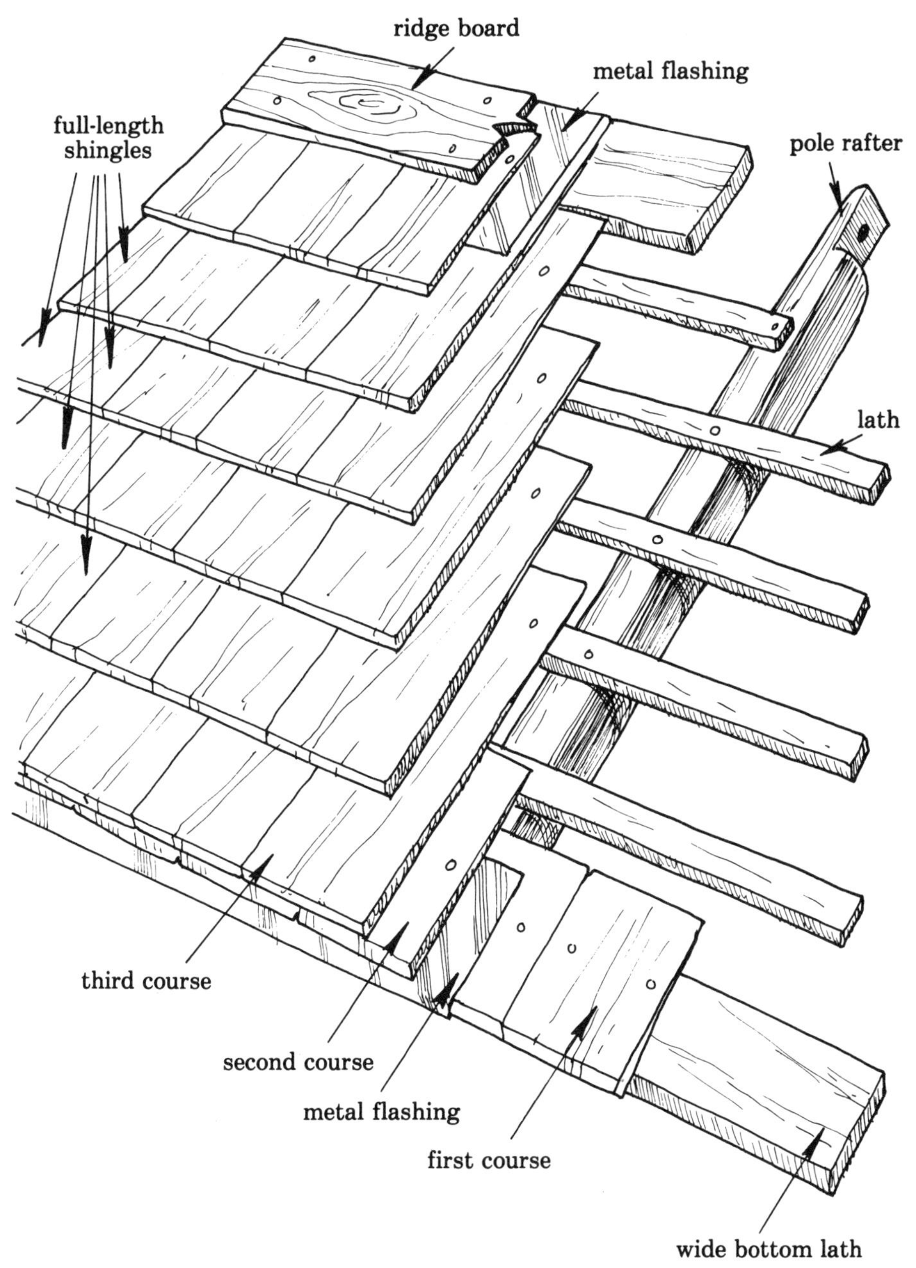

Illus. 10.6. ***Three-course shingling (on noninsulated roof).***

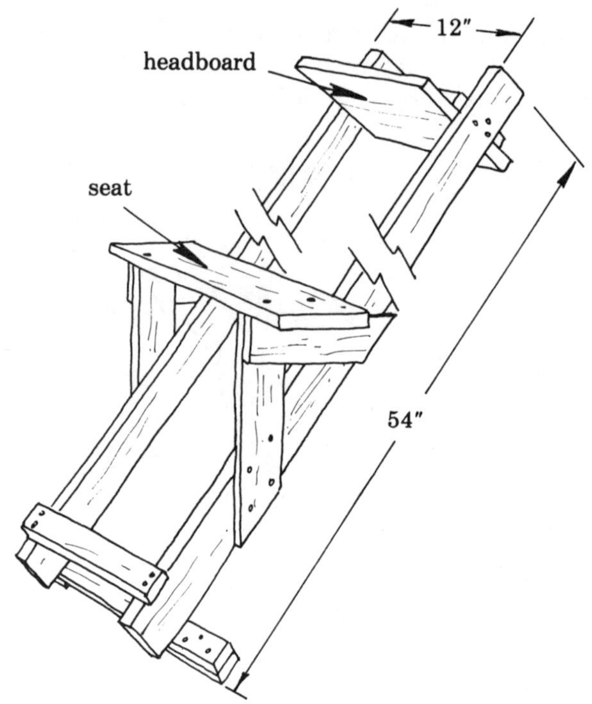

Illus. 10.7. *Shingling seat. The headboard hooks over roofing lath or ridge. Seat provides level work area.*

Photo 10.7 *Furring strips on this roof were nailed in place at the same time shingles were applied. Almost all work was done from scaffolds inside.*

On the other hand, the combination of logs and shingles is extremely attractive and shingles can be a pleasure to make—if you have the time and suitable timber. Shingles are also one of the few roofing materials that you can make with simple hand tools.

Shingles require high-quality, straight-grained wood and a plentiful supply of nails. In the past, many areas had plenty of quality trees available, but nails were scarce. Japanese housewrights solved the problem by pegging bamboo dowels through the exposed portion of very thin ($^1/_{10}$ inch) pine shingles.

Nowadays nails are plentiful, but time seems scarce and good shingle wood is hard to locate. I can make about 200 shingles in a day, and our roof required about 5,000 shingles, or 25 full days' splitting and dressing. We also went through a 50-pound box of nails. As for the wood, though we live in one of the best hardwood areas in North America, locating suitable timber was a challenge even here.

We needed four logs for the shingles: two 12-foot logs an average of 18 inches in diameter, one 9-foot veneer-grade log 29 inches in diameter and one 16-foot log about 18 inches in diameter. Two possible sources are

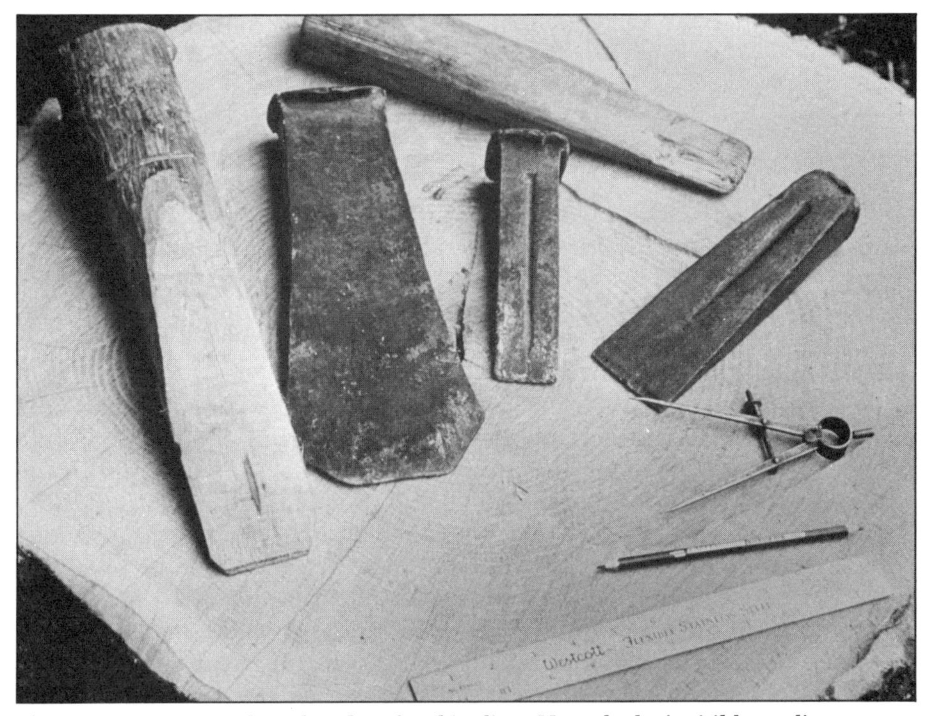

Photo 10.8 ***Layout tools and wedges for shingling. Note the log's visible ray lines.***

sawmills or log brokers. The advantage of purchasing logs (compared with cutting your own) is that you can choose from a wide selection. Shingles are split only from the butt cut of the highest-grade logs. At a sawmill you can examine the end grain of all the logs. Since logs are generally bucked 6 inches longer than final board length, many sawyers won't mind if you resaw a 1-inch round from one end for a fresh look inside. But they'll be charging top prices for select cuts: I paid $200.00 to $400.00 per thousand board feet for red oak logs in 1979. The four logs cost a total of $285.00—or $31.65 per 100 square feet (comparable to the cost of asphalt shingles, not counting my labor). The cost per shingle came to 5.7¢.

It's hard to know how long a shingle roof will last. Some pine shingles used as siding in New England have lasted over 300 years, but we demolished a house built in the 1930s, and the pine-shingle roofing was in tatters. In our area it is said that chestnut shingles were good for 50 years, white oak for 40 years and red oak for about 20 years—this is with no preservative treatments. In North America the list of coniferous shingle trees includes pines, cypress, western red cedar, some firs and balsam. Hardwood shingles are usually white or red oak. I've also heard of tulip poplar being used for shingles on temporary structures.

The thickness of the shingles can vary from 1/10 to 3/4 inch, with the thinner ones applied at least three deep. The best shingles are tapered in thickness, because the thick portion, exposed to the weather, deteriorates much faster than the thin, covered portion. The taper also makes it easier to lay the overlapping courses. Shingles can be from 3 to 8 inches wide.

CHOOSING GOOD SHINGLE WOOD

In general, large-diameter logs are superior and represent a better value than small logs, because there's more waste with small ones. The width of the finished shingles is also determined by the diameter of the log used. Roughly, a 16-inch-diameter log should make shingles 3 to 4 inches wide. A good 24-inch log makes 5- to 7-inch shingles. With a large log you get more shingles with less work.

Look for a straight, round trunk with no visible knots. A small knot may mean grain distortion on each side for a space equal to the knot diameter. Examine the furrows in the bark. A twisting pattern indicates spiral growth of the tree. This results in parabolic shingles. Look carefully for knot scars. The bark will carry an indication of interior knots many years after a limb falls off. Use an axe to peel off a strip of bark.

Examination of the end grain at the butt and tip gives some indication of the inner log. If the log looks good, saw an inch off one end. You'll get a better idea of wood quality by looking at freshly exposed end grain. Check the thickness of the sapwood zone. Sapwood has to be discarded as it rots much faster than heartwood (except with cedar). Chop the round into small radial pieces to see how it splits. Good shingle wood splits very easily.

Good wood is sound, without any insect holes or discoloration that could indicate rot.

When I find a good log, I measure it from the tip end into 20-inch increments. I start at the tip because that puts any waste (less than 20 inches) at the butt end, which is more likely to contain uneven grain. Saw the log at right angles to its length. Crooked cuts result in shingles of uneven length.

SPLITTING TOOLS

You'll need a sledgehammer, several wooden wedges, an axe, a froe, a club and a brake. Dressing requires a broad hatchet, a shaving horse and a drawknife. I also use a colored pencil, a ruler, and a pair of straight-leg dividers.

A shingle froe has an 8- to 12-inch-long horizontal blade and a vertical handle. You strike the froe into a bolt of wood and twist the handle to break off the shingles. Larger froes are awkward to use. The blade should be $5/16$ to $7/16$ inch thick and rounded across the striking edge so that the club doesn't get battered too quickly. The splitting edge should be beveled gradually; too short a bevel is hard to drive into the wood. For good leverage, froe handles should be 14 to 18 inches long; any straight-grained hardwood will make a good handle. Some froes are made with a tapered eye, like a pick axe. The iron won't fall off, but it tends to come loose. Wedges driven into the wood to secure the froe blade don't hold up to the constant pounding of the club either. I prefer a parallel-side socket. The handle should be fitted slightly oversized when the wood is oven dry, then the handle will swell and tighten after installation, when moisture from the air reenters the handle. The iron can be secured with a lag bolt and a large washer.

I use a long, narrow club to drive the froe. The small diameter makes it easy to strike the froe once it's driven into a bolt of wood. The weight comes from the club's length. Any dense hardwood limb or sapling can be used for the club. Froe clubs should be allowed to season before use. Incorporating a knot (or a cluster of knots) in the striking end will greatly increase durability. Froe clubs are easy to make and are expendable. I destroyed a good hickory club about halfway through our 5,000 shingles.

A brake consists of two horizontal lengths of wood into which the bolt is wedged—securing it while it's being split (see photo 10.9). The makeshift brake that I use is made of 2 × 4s nailed to a very large oak stump. The nice thing about this setup is that the end grain of the stump is used as a work surface to support the shingle bolt while the froe is being positioned. Set the two horizontal members about 4 inches apart.

SPLITTING THE SHINGLES

Usually there is at least an incipient crack or two originating at the pith of the log to be split. It's a good idea to halve the log along this

Photo 10.9 ***Splitting the bolts with a maul and froe at a shingle brake.***

original crack. Begin by penciling an extension of the crack along a ray line, outward to the sapwood. Start the first split with a flat iron wedge placed on the line, at the edge of the sapwood. Strike the wedge, allowing the log to crack naturally. Use iron splitting wedges to open the crack, followed by a wide hardwood wedge (called a glut) to finish the splitting and release the iron wedges.

Photo 10.10 *Marking out 3½-inch sections along the ray lines, beginning at a natural crack.*

Photo 10.11 *Starting the first split in a crack.*

Photo 10.12 *Driving in a wedge to allow the log to begin cracking on the opposite side.*

Photo 10.13 *If possible, split the log into sections consisting of four 3½-inch bolts.*

Photo 10.14 *On logs large enough to yield an inner ring of shingles, mark inner bolt dimensions along the annual rings at 3½-inch widths.*

Photo 10.15 ***Splitting bolts parallel to the annual rings.***

Photo 10.16 ***Splitting bolts from the outer ring.***

Photo 10.17 ***Splitting off the sapwood.***

Photo 10.18 ***Splitting off the inner heartwood, plus uneven wood and knots.***

Photo 10.19 ***Separating the bolts. If necessary, use an axe to cut through unsplit connecting fibers.***

Split each half-log into 3½-inch segments, then split off the sapwood and inner heartwood from each to form trapezoid-shaped bolts. Place the froe along the center of the bolt. Careful centering is critical. If the froe is off to either side, the split will probably run out. Strike the froe directly above the bolt. Place the bolt with the embedded froe between the legs of the brake and continue opening the split by pulling down on the froe handle. If necessary, use the club to drive the froe as the split opens.

There's not much you can do to control the first splitting of a 3½-inch bolt. It usually breaks quickly, almost snapping apart. Split the resulting 1¾-inch-thick segments in the same manner to ⅞-inch pieces. These in turn are split to the final 7/16-inch thickness. With the thinner divisions, the froe can be used to influence the thickness as the split is made. The trick is to apply pressure against the thicker half of the segment, correcting the division by controlled cross tearing. Place the segment in the brake with the thicker section facing the near leg. Pull the froe handle down against the thick section, if possible, simultaneously placing your

Photo 10.20 ***Splitting a ⅞-inch quarter-bolt into 7/16-inch shingles.***

free hand into the split and pulling down. The direction of the split may correct quickly and suddenly begin running out on the other side. If this happens, rotate the piece and correct the split again, this time from the other side.

It's sometimes difficult to position the froe at exact center, especially if there's a jagged saw cut or a slight bump on the end grain. If necessary, use a broad hatchet to trim a flat or concave surface so the froe can be positioned properly. Don't attempt to correct a crooked split by relocating the froe. You must continue with a split once it's started.

DRESSING

Most split shingles require edge trimming and side dressing. Shingling is similar to masonry in that each course overlaps joints in the previous layer, at the same time setting the stage for the next row. Work goes much faster with flat, uniformly thick shingles, and the resulting surface will last longer. The edges of split-oak shingles are often wavy, often with some sapwood left on the outer edge. The finished edges must be straight and parallel and all sapwood must be removed. Use a broad hatchet, a smaller version of a broadaxe with the edge beveled on only one side of the blade. The inner side of the blade should be flat from the poll to the edge. The bevel must be thin, about 25 degrees, and sharp. Edge trimming is done at a clean, hip-high hardwood stump with a small indentation in the surface to keep the shingles from slipping.

Face dressing is necessary if the shingles are too thick, or if you want to give them a tapered profile. It's easier to lay shingles if the overlapped tips are a uniform thickness, and this means dressing the thicker shingles. I use a drawknife with a slightly curved blade, the beveled edge facing up. Keeping the proper shape requires some extra attention, since the concave bevel can't be dressed with a flat stone. However, the slight curve is excellent for taking wide shavings off an already flat surface.

First determine which end of the shingle will be exposed to the weather. If it's too thick, clamp the tip into a shaving horse and shave the butt to the specified thickness (7/16 inch in our case). Then reverse the shingle and shave a long, thin taper toward the tip. Our ideal shingles were 3/16 inch thick at the tip ends. Check for warp in the plane or irregularities between the two sides. There is often a choice of which side to shave, giving an opportunity to correct slightly twisted grain or uneven splitting.

After dressing, tie the shingles in bundles. Put 20 shingles into a wood vise to compress them, then tie them into a bundle with two loops of garden twine. Bundling restrains warpage and makes the shingles easy to carry. We then treated our shingles with a water repellent as described in chapter 12, Finish Work. We dipped the bundles before nailing, and followed up with a brush-on coat the next year. We intend to re-treat the shingles every ten years.

Photo 10.21 *Side dressing a shingle with a broad hatchet.*

Photo 10.22 *Face dressing shingles at a shaving horse.*

Photo 10.23 ***A slightly curved drawknife, used bevel up, removes wide, thin shavings.***

INSTALLING A SHINGLE ROOF

On modern insulated roofs shingles are often applied two layers deep with heavy roofing felt under each course, but we nailed our shingles three layers deep without paper. The added air circulation around the shingles helps prevent decay and minimizes warpage. For this to work, the attic must be well ventilated, and the shingles should be nailed to spaced laths, not to solid sheathing. A minimum 8-on-12 pitch is recommended for fast drainage.

Begin the bottom course with a row of 8-inch-long shingles, with a strip of flashing tacked 1 inch below the shingle butts as a drip edge. The second layer uses shingles 14 inches long, nailed over the first row, with a common lower edge. The third layer uses full-length (20-inch) shingles, also lined up at the butts. The varied lengths form a three-tier, tapered base so that successive courses lie flat against the lath. The remaining courses are each spaced to leave 6 inches of shingle exposed to the weather. Shingles are nailed side by side in random widths. To keep the rows even, snap a chalk line over the previous tier of shingles, 6 inches up from its lower edge, where you intend the bottom of the next course to lie.

Photo 10.24 *A shingling seat is a big help.*

Photo 10.25 *Finishing the ridge. The platform visible below the gable allows access into the attic.*

Use two 6d (sixpenny) galvanized roofing nails for each shingle. Locate nails so they are covered by following layers of shingles. Our 20-inch shingles were nailed 7 to 8 inches above the lower edge and ¾ to 1 inch from each side. A gap between shingles allows expansion to occur harmlessly. This space can vary from 3/16 to ⅜ inch, according to the size and kind of wood used. Softwood shingles require larger spacing than hardwoods. Wide shingles expand more than narrow ones and require more space. Treated shingles should require less spacing.

Sometimes oak shingles split during nailing. Splitting can often be prevented by firmly pressing down on the shingle with the left hand during nailing to minimize vibration. If a shingle begins to split around a nail, it must be removed. Slightly warped shingles can sometimes be nailed flat, but poor-fitting shingles must be discarded. Nail cupped shingles concave side downward.

Finish the ridge as you did the first course, with three overlapping layers of shingles. This time, though, each layer should be made of shingles that are shorter, not longer, than those in the layer below. Attach a strip of flashing along the ridge, over the ends of the next-to-last row of shingles. The last row of shingles can be nailed horizontally along the ridge, either overlapped like the vertical shingles or placed end to end with no overlap.

CHAPTER XI

CHINKING

Chinking can be one of the most attractive features of a hand-hewn log building. The smooth, wavy bands are a perfect complement for the flat, textured logs. But the filled chinks can be the weakest link in hewn-log construction. Chinking may crack, or even fall loose, in some cases requiring yearly maintenance. It takes time to do a good chinking job. It may take one person two to three days to insulate and chink both sides of one 8 by 20-foot wall. Some logbuilders hold chinking parties to speed up the work.

Photo 11.1 *Stucco (cement mortar) chinking.*

Although a great deal can be accomplished quickly that way, quality control is difficult, and cleanup may be a big job.

Cracks caused by the differing expansion rates of logs and the mortar fill show up as horizontal gaps between the two. Most log shrinkage occurs during the first year after hewing, but even seasoned wood will continue to expand and contract indefinitely as its moisture content seeks to parallel changes in atmospheric humidity. In order to minimize horizontal cracking between logs and fill, put off chinking until the year after hewing. This waiting period should include several months for the notches to settle after the log walls are raised. Try to chink during dry weather.

Waiting for logs to season may seem inconvenient. However, most hand-hewn log structures take well over one year to build. Ideally, you could hew logs during winter (wonderful weather for axework), then notch and raise the pen in spring. Roofing, insulation, plumbing, wiring, and subflooring would then be put in during the summer months. By fall, most log shrinkage would have taken place, and the notches would be well settled. Don't install the door and window frames until just before the chinking is applied, otherwise log shrinkage could cause their frames to buckle or logs to shift out of position. Dry fall weather is an excellent time to chink a wall—the cool days are good for curing the chinking.

Vertical cracks are caused by uneven expansion of the chinking materials. Ted Goudvis, an engineer who pioneered prestressed concrete construction methods, believes that the extremely small size of clay particles used in traditional formulas is responsible for much of the vertical cracking in them. Ted recommends using standard stucco (Portland cement mortar) as a crack-resistant substitute.

TRADITIONAL METHODS

Although many methods of filling the gaps in a log house have been used over the years, most of them start with short wedges of seasoned firewood (or small stones) placed over dry moss. The moss fills the middle of the gap, and the wood wedges or stones serve as fill. Finally, nails hammered into the edges of the gap serve as a base for the mortarlike top coat.

A wide range of chinking formulas has been advocated. One old recipe calls for 4 parts clay, 2 parts ashes, and 1 part salt. I've also heard of using lime-sand mortar, plain clay, clay mixed with chopped straw, and cow manure mixed with chopped straw. (The latter is supposed to be excellent.)

Daniel O'Hagan's chinking formula consists of 1 part clay, 3 parts sand, 1 part ashes, 1 part lime and some chopped straw. Daniel recommends yearly maintenance using a brush-on paste of white cement for dressing cracks. He uses white cement because it reflects light well and so helps to keep the interior brighter.

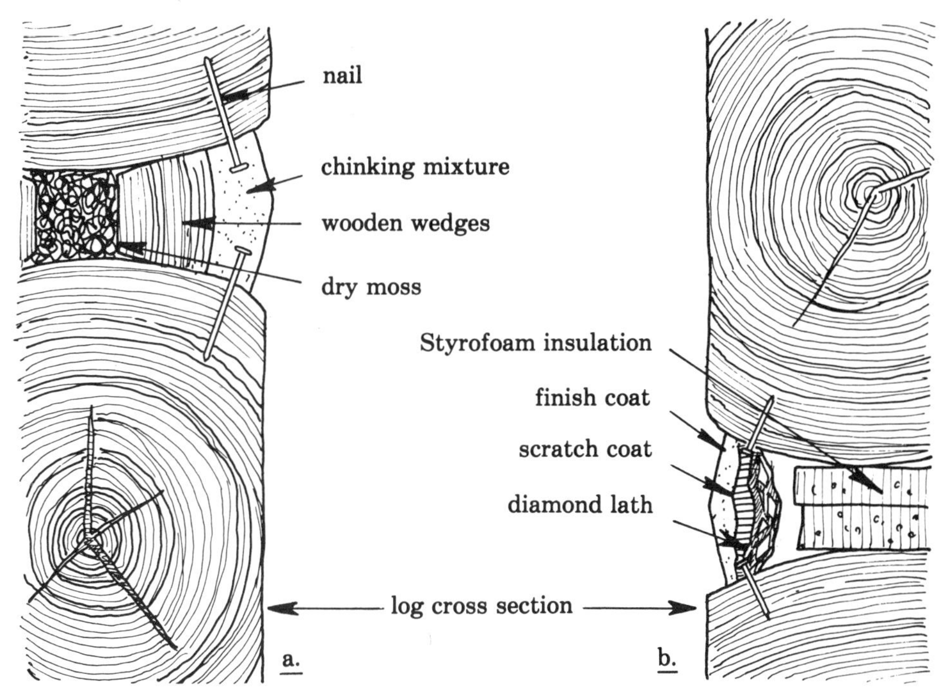

Illus. 11.1. ***(a) Traditional chinking. (b) Modern chinking. Use stucco mortar for the scratch and finish coats.***

When Peter Gott experimented with chinking mixtures, he came up with a formula based on a ratio of 1 part cement to 3 parts other solids. Peter uses 2 parts cement, 3 parts clay and 3 parts sand. The clay gives this mixture a good working consistency and a nice, natural color, but the ingredients must be mixed very carefully.

HOW TO CHINK

The problems often associated with log chinking have been solved in recent years with a composite chinking system consisting of stucco (cement mortar) around a commercial insulation such as Styrofoam. The insulated core gives an R-value higher than that of the logs. Stucco doesn't crack, although small horizontal gaps between logs and mortar may open during dry weather. The foam core should still stop air infiltration.

First put in a lath backing to which the stucco can adhere, starting with either side of the log wall. When you've stuccoed one side, insert insulation from the other and repeat the stuccoing on that side, working with different parts of the wall in different stages, so there's always something to do without having to wait.

Plasterer's metal diamond lath works well (diamond refers to the shape of holes formed by the lath wire). Diamond lath is sold in 2 by 8-foot sheets. You can also use other types of metal lath, ½-inch hardware cloth, or several layers of poultry netting. Diamond lath may be difficult to locate. If necessary, ask a stucco and plaster contractor about a source. We used 15 sheets to chink our two-story, 16 by 20-foot house.

With tin snips cut the lath into strips about ¼ inch narrower than the width of the chinks at the log faces (where the hewing starts). Diamond lath is nasty stuff to work with, especially after frazzled fingertips touch the mortar, which stings on contact with cuts. Wear gloves whenever possible. Lengthwise strips are easier to bend than strips cut across the diamond pattern.

Curl the strips slightly, then wedge them about an inch into the gap. Tack the upper and lower margins to the logs, using short, wide-headed tacks sold for stucco lath. Space the tacks every 6 inches. Use a wide nail set (such as a ½-inch bolt) to start tacks in tight corners.

SCRATCH COAT

Stucco is applied in two layers. The first, called the scratch coat, adheres to the metal lath and provides a rough bonding surface for the

Photo 11.2 ***Metal lath tacked in place. Styrofoam insulation has been fitted from the other side.***

second, called the finish coat. Use 1 part bagged mortar mix to 3 parts clean bricklayer's sand for the scratch coat. Mix the ingredients thoroughly before adding water. You may not notice sand pockets in the mortar until pieces of chinking begin falling loose from your house. Add clean water. The scratch coat can be a rather stiff mix. Hand mix small batches in a mortar boat. Make what you can use up in about two hours.

Chink the upper logs first. Use portable scaffolding and haul mortar in a bucket (a piece of garden hose, slit and attached to the bucket handles helps). Use a mortar hawk or a large plasterer's trowel to hold small quantities of mortar as you work. Margin trowels are an ideal shape for applying stucco to the long, narrow gaps. Odd-shaped corner areas require a small pointing trowel. Carefully clean mortar tools before taking breaks, otherwise the stuff will set hard and you'll never get it off.

The scratch coat should be about ½ inch thick. The face can be left flat, or troweled to a slightly convex shape. The edges of the base coat

Photo 11.3 *Scratch coat of cement mortar forms a base for the tinted finish coat.*

Photo 11.4 ***Portable hanging scaffolds are very useful.***

should be kept at about ½ inch in from the curved faces of the logs. Leave the surface rough and unfinished. If mortar in the bucket begins to set, remoisten and remix it once. Cure the scratch coat for several days before adding the finish coat. As with other mortar, exposure to direct sun should be avoided. Spray on water when the stucco seems too dry.

FINISH COAT

A fair amount of skill is required in order to apply a smooth finish coat. Practice on areas that will be hidden by cabinets and furniture. The scratch coat must be dampened (but not flooded) with water. The mortar consistency should resemble a thick milkshake, more liquid than the scratch coat. The cross-sectional view of the finish coat can be flat, rounded or slightly V-shaped. With either style, provide a drip edge by keeping the upper edge about ¼ inch inside the wall surface. The V-shape is not difficult to make, and I like the way it looks. Trowel the lower half smooth, running the edge of the trowel along the curved surface of the lower log. Keep the trowel tilted slightly outward, at the angle required for the V. Keep the mortar about ¼ inch in from the log sides. Do the upper half the same way. If necessary, touch up the lower half lightly. Learn to trowel with long, smooth strokes.

The dull gray cast of cement can be modified with a small amount of colorant added to the finish coat. Powdered pigments are available from cement contracting suppliers. Experiment with the colorant by adding various amounts to batches of scratch coat mortar. Allow the colors to

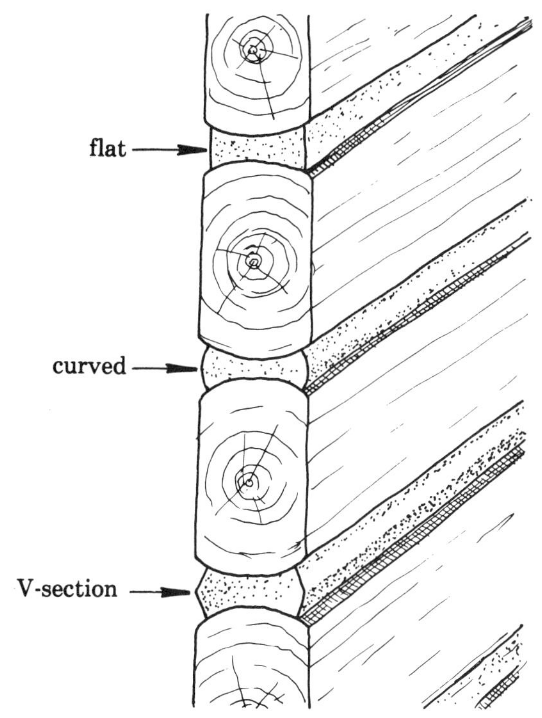

Illus. 11.2. *Chinking finish styles.*

change completely by letting the mortar fully cure. We used 2 ounces of ochre in a batch consisting of 3 shovels of mortar mix and 9 shovels of sand.

INSULATION

Insulation should be put in while the mortar scratch coat is curing on the opposite side of the wall. Styrofoam panels can be cut into strips 3 to 4 inches wide and snugly tucked into the center area. Fill the area between the logs very carefully, cutting pieces to fit odd spaces. The insulation should completely fill an area at least 3 inches wide with the strips placed flat side down.

Sprayed urethane foam can be used instead of Styrofoam. It completely seals cracks and has the highest R-value of any commercial insulation, but it must be carefully applied since it can cause severe allergic reactions if not completely sealed within the wall. Strips of fiberglass batts can also be used, but fiberglass won't seal drafts unless it is tightly compressed, in which case the insulation value decreases greatly.

Polystyrene rope, sometimes called backup rod, is manufactured in ½- and 1-inch diameters. The ½-inch size is excellent for filling odd gaps.

CAULKING

Small cracks that leak air should be filled with caulking. Typical leaky areas include the interface where door and window frames join with logs and chinking, notch corners, and places where joists are mortised through wall logs. You may also want to caulk exterior log checks to keep moisture out.

Neoprene caulking is excellent, though it's expensive and somewhat difficult to work with. I use latex caulking, which is available in various colors. Bronze seems to be the least conspicuous. Check into prices by the case. I found them to range widely.

Caulking must be applied during warm weather on clean, dry surfaces. If necessary, the tubes should be prewarmed. Push the caulking bead forward with the tube pointed toward the uncaulked area. (The natural inclination is to pull the bead, but this won't fill joints as well.) The bead should be concave, if possible, with a minimum depth equal to the width of the opening. Gaps over ¼ inch wide should be backed with polystyrene rope.

CHAPTER XII

FINISH WORK

Finish work may seem both frustrating and satisfying at the same time. The fact is, it's hard to keep up the momentum at this stage, and for many homebuilders, finish work can drag on for years, a constant irritant that detracts from the joy of having done it yourself. Even if you have to reduce your involvement at this stage, don't let progress come to a standstill. Take the jobs one at a time, and let steady progress sustain your commitment to see it through, down to the last strip of molding.

LUMBER FOR FINISHING

It's surprising how much milled lumber can be used in building a log house. A list of basic finish work includes subflooring and finish flooring, door and window framing, doors, sashes and stairs. Many people who build with logs have access to a woodlot, where trees can be selected and taken to a sawmill for milling. Or you can buy logs for finish wood and have them milled, avoiding the expense of buying finished boards. Many of the wood species used for log construction are also suitable for lumber. You can set aside logs for this or use butt ends that contain wavy grain unsuitable for hewing. And knotty logs that turn out to be too hard to hew can still be used for shelving, paneling and cabinetwork. Edge lumber and poorer-grade logs can be sawed into shingle lath, gable sheathing, 2 × 4s, insulation strapping or foundation forms. (The form lumber can be recycled as temporary flooring or scaffold planks during construction.)

Sawmills generally saw boards into thicknesses based on ¼-inch increments. A 1-inch board is specified 4/4 (four quarters). Lumber to be dressed (planed) should be sawed an extra ¼ inch thick and ½ inch wide to allow for the amount to be cut off by the planer. (A high-speed, single-pass planer will dress a true 1 × 6 into a ¾ by 5½-inch board.)

All lumber should be sorted and carefully stacked on spacers (called stickers) to prevent mold and fungus development, warping or insect invasion. Under good conditions, most 1-inch boards will air dry to 20 percent moisture content within two or three months.

Air-dried lumber can be used for door and window frames, flooring or staircases. However, the wood will shrink during the dry winter months. Stairways in particular—because of their great amount of wood—will

Photo 12.1 ***Door and timber-framed overhang at the Gotts'.***

shrink. Because of this, it's a good idea not to attach them permanently to log walls or joists for at least two years after the house has been closed in. Ideally, sashes for windows and doors, and doors themselves, should be dried further to a moisture content similar to wood in an occupied house during the winter. The recommended moisture content of wood to be used for interiors in most of the United States is 8 percent. The level for the humid Gulf states and the southeastern seaboard is 11 percent. In the arid Southwest, moisture content should be brought down to 6 percent.

Drying can be accomplished by stacking stickered lumber indoors where there is a heater, perhaps by closing off part of the house or a section of a barn or shed, then setting up a slow-burning woodstove. Use an electronic moisture meter to test the moisture content, or get an approximate measurement by comparing the weight of a small oven-dried sample to a sample of the same size from the bulk stack. Dry lumber should be used immediately or wrapped tightly in plastic sheets—otherwise the moisture content will soon return to the level of air-dried wood.

HOMEMADE FLOORING

When visitors step into our house for the first time, they often acclaim the beauty of our tulip poplar flooring. The floors aren't perfect, but they are attractive and should take many years of hard use. The wood was milled from some of the same trees that provided our wall logs, at a small fraction of the cost of the commercial wood flooring. Hardwoods suitable for floors include red oak, white oak, birch, maple and beech. Pine and other softwoods aren't as durable but can be just as attractive. Wide pine planking was a traditional favorite in New England.

Floors are generally installed after the roofing and chinking. The subfloor boards should be seasoned and dressed (planed on four sides), because uniform boards are essential to provide a secure base for the finish flooring. Full 1 × 6s (planed from boards 5/4 by 6½ inches) make excellent subflooring. Random widths can be used, though they take longer to install. You can also purchase inexpensive 4 by 8-foot sheets of subfloor material. Green (unseasoned) subflooring will also work, but you'll get gaps in winter as wide as ¼ inch between 1 × 6 boards. Try to lay flooring in dry weather to minimize shrinkage.

If at all possible, dry the finish flooring artificially, then cut in tongue and grooves or half-laps. These overlaps seal drafts and keep dust from falling through. Tightly fitted tongue-and-groove flooring will add considerably to the rigidity of the floor and the structure in general. You can add a bevel to the edges to reduce the visual effect of any spaces between the boards. I put a ⅛-inch bevel on the upper edges of our floorboards with a hand plane.

Single-layer flooring (no subfloor) can be used between first and second floors. Tongue-and-groove flooring dressed to 6/4 is recommended. Our tongue-and-groove poplar is 25/32 inch—a common finish floor thickness that is really too thin for single-layer use.

Floor joists made of logs usually require some leveling before the subfloor can be laid. You can dress the joist tops with a lightweight, flat adze. Use a straight wood strip to locate high points, and mark them with a pencil. It's not necessary to level all the joists down to the lowest point, but be sure that about half the width of each joist will make contact with the flooring. Use shims on downstairs joists above the crawl space to help provide a firm surface for the flooring.

Subflooring, softwood finish flooring and poplar should be face-nailed, but hardwood tongue-and-groove boards are nailed through the tongues at an angle. Use 7d or 8d screw-type flooring nails for all flooring

$^{25}/_{32}$ inch or thicker. Countersink any face-nailed flooring, and fill the nail holes with putty. For strength and appearance, joints of adjacent boards should never fall in the same place. Also, avoid overlapping the joints of finish flooring on subfloor joints. Laying the subfloor boards diagonally strengthens the structure and avoids the problem of overlaps.

Tight nailing may require that you force some boards to straighten them out. I sometimes nail a cleat into the joist, then insert a wedge that forces the flooring tightly in place. For the wedge, use the groove side of a scrap piece of flooring. You can also drive a chisel into the joist, then pry back against the floorboard.

We rented a floor sander, powered by a portable generator, to smooth the floors. Then we sealed and varnished the boards with three coats of satin polyurethane, hand sanding very lightly between coats.

BUCKS AND LINTELS

Door and window frames for log houses are sometimes called bucks. Bucks for hewn- and chinked-log structures generally consist of heavy lumber spiked directly into the end grain of wall logs, although with chinkless log construction (where frames do not directly support log weight), special floating frames can be built. The bucks serve the double purposes of supporting doors and window sashes and securing cut-off log sections. Since logs shrink in diameter, bucks should not be installed until most log shrinkage has taken place. The waiting time should be at least one year. Install the bucks just before chinking.

Lintels are the supporting beams above window or door openings, and in log houses, they are usually regular full-length wall logs with space cut out to accommodate the door or window frames. These upper cutouts for doors and windows should be made at the same time that corner notches are executed. In most cases, the full-length log is notched to accommodate the door or window frame. The whole job can be done at once, with the logs turned upside down and worked on trestles. Overhead notching is extremely difficult. Lintels can be square, arched or double curved, but the cutout depth should not exceed half of the log diameter.

Log openings must be trimmed before setting the bucks in place, and the sill notches should be cut at the same time. The sill cutouts should not be made in advance, since it's impossible to determine their exact depth until you have seen how much log shrinkage occurs. When you lay out the sides of the opening for the rough cut, use a long level to outline the cutouts on both sides of the log wall. Take measurements from a reliable reference, such as the vertical cut at the notches, and use a short, two-man crosscut saw to even out the sides of the opening. If you use a chain saw, make a guide fence by nailing a square 2 × 4 to the side of the opening outline. The sill notch is cut much like any other notch with a level notch

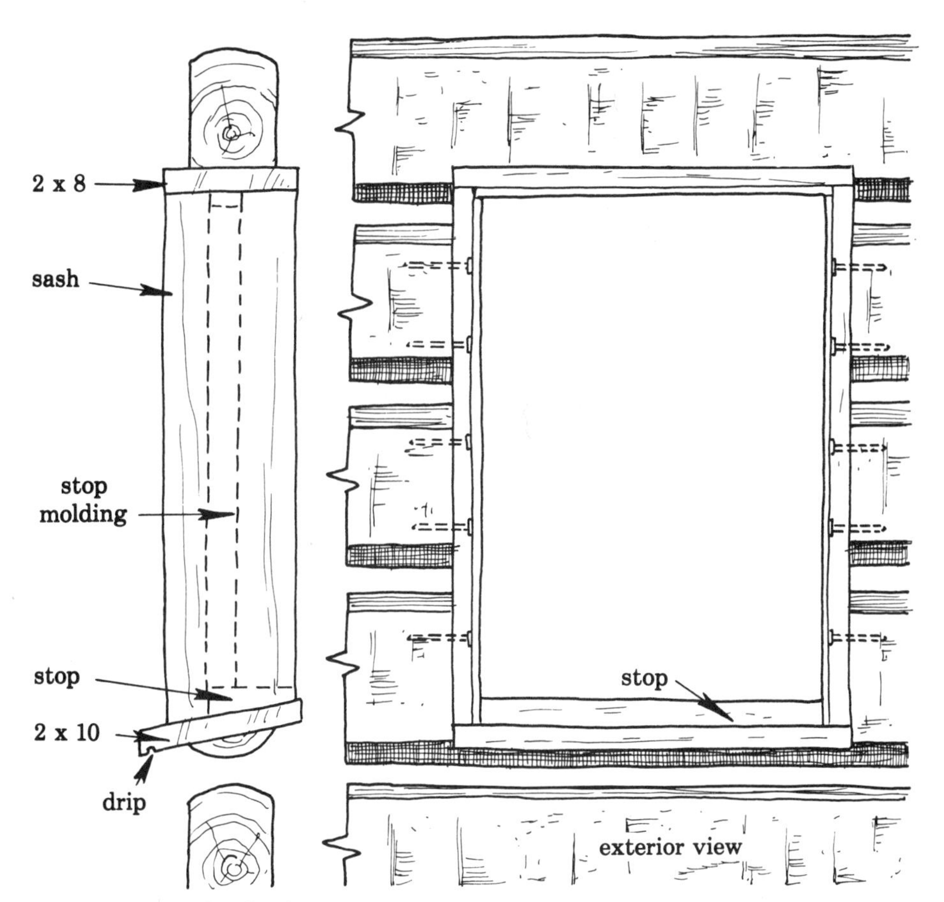

Illus. 12.1. *Window buck.*

face. Saw a row of kerfs to a chalked, level, bottom line, and knock out the waste with a chisel and mallet. Pare the surface smooth.

Door and window bucks are very similar, but the doorsill should be of hardwood for durability. Windowsills are usually sloped to drain off rainwater; this requires a sloped notch. The head and jambs (sides) can be 2 × 8s. The sills, which overhang the logs somewhat, should be 2 × 10s. The lumber should be dressed "four square," that is, square at all corners.

Careful planning is necessary to ensure that doors and windows fit and function properly. Standardizing their opening widths and heights simplifies construction. Saw cuts should be carefully laid out and executed. After nailing, square each unit, then tack temporary diagonal braces across the frames. Be sure that the jambs are set plumb. Gaps between log cutouts

Photo 12.2 *A simple window buck: four 2-inch-thick planks butt-joined and nailed together.*

Photo 12.3 *Lintel cutouts are notched with log upside down.*

and bucks should be filled with tapered wedges (I use shingles) before nailing or doweling, wherever the nail or dowel will penetrate. Hammer two 6-inch spikes through the side jambs into the end grain of each log, but don't nail the sills and head jambs to the wall logs (log shrinkage could pull the bucks apart). Locate the spikes so that the nail heads will be covered by interior molding. Treat bucks with a water repellent before installing them.

DOORS

Doors can warp, swell or shrink as their moisture content changes from day to day and season to season. The result can be doors that stick at certain times of the year and admit outside air at others (usually the coldest months). Door wood should thus be dried to the same moisture level that interior wood in the house will have during the winter months. Exterior doors should be carefully weather-sealed after installation with weather strips around the edges and several coats of varnish. The varnish will reduce swelling by limiting the infiltration of moisture into the wood.

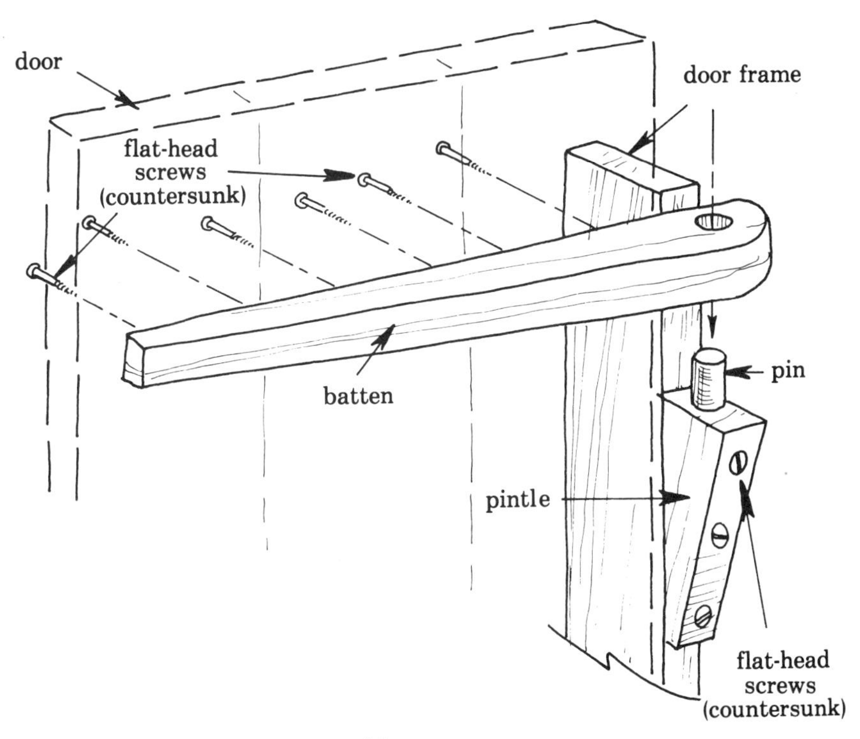

Illus. 12.2. *Wooden hinge assembly.*

Using dry wood in the first place will help prevent the dry interior air in winter from drying out the door, causing it to shrink.

Single-thickness planks are suitable for interior doors, but a minimum of two layers of boards should be used for exterior doors to give them some insulation value. The best insulators, of course, are hollow-core doors filled with foam insulation, but the main criterion for doors is careful installation, cutting air infiltration to a minimum.

The single-plank designs that I use are straightforward and harmonious with log construction. While these doors require careful craftsmanship, only basic hand tools are required to make them. Our doors are built with two or three wide planks, originally sawed 1¼ inches thick. I planed the doors to 1⅛ inches in thickness. I used hardwood hinges, which also serve as cross battens.

The boards are joined by hidden doweled patches mortised into the adjoining edges (see illus. 12.3). Each edge also has a groove along the full length of the board that is filled with a spline (except at the location of the

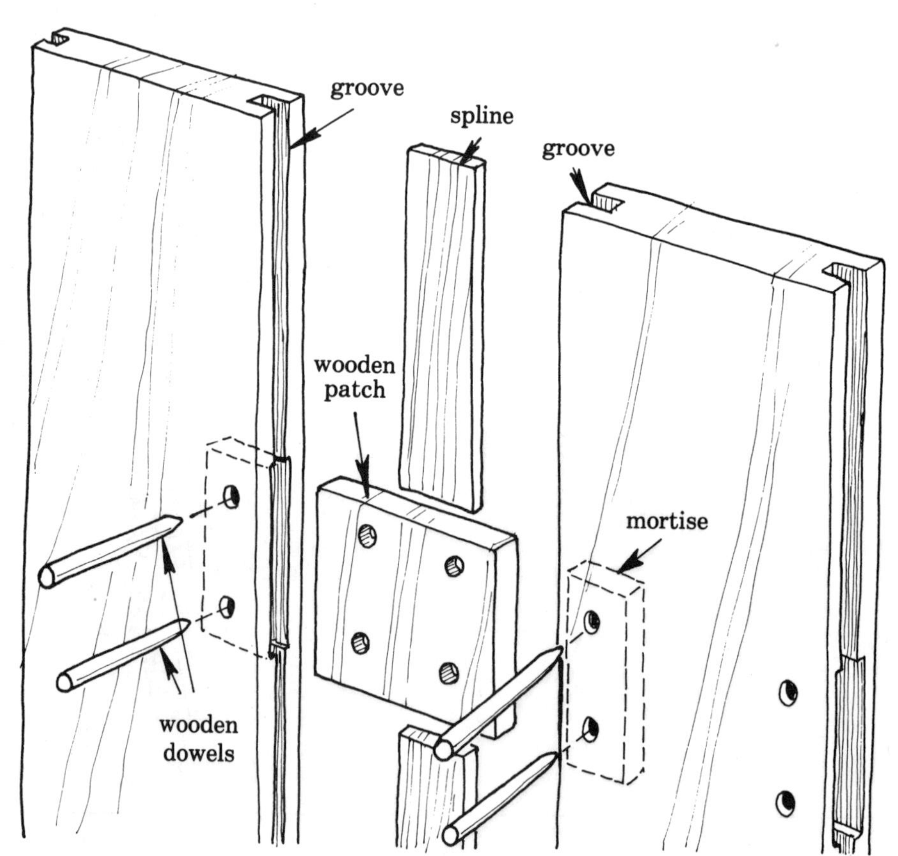

Illus. 12.3. ***Hidden patch-and-spline detail.***

patches) twice as wide as the groove. It thus extends into the groove in the adjacent board, weatherproofing the joint. The patches are small hardwood squares, about 4 by 4 by 3/8 inches. Chisel two mortises half the size of the patches into each door edge. Space the patches about one-quarter of the door height from the top and bottom of the boards. Bore two 3/8-inch holes through the area of each board where the patch will go, and preassemble the boards with the patches in place. Mark the centers of the holes on the patches. Disassemble the boards, bore holes in the patches 1/8 inch closer to the patch center than the mark, and reassemble the boards with patches and splines.

The slight off-center placement of the patch holes pulls the boards toward each other when the dowels are driven through, resulting in a tight joint that does not require glue. Use pointed, 3/8-inch dowels, longer than the board is thick, and hammer them through the holes. The dowels can be trimmed flush to the board surface or left to protrude 1/4 inch or so beyond the surface. If you use dowels as fillers for countersunk screws

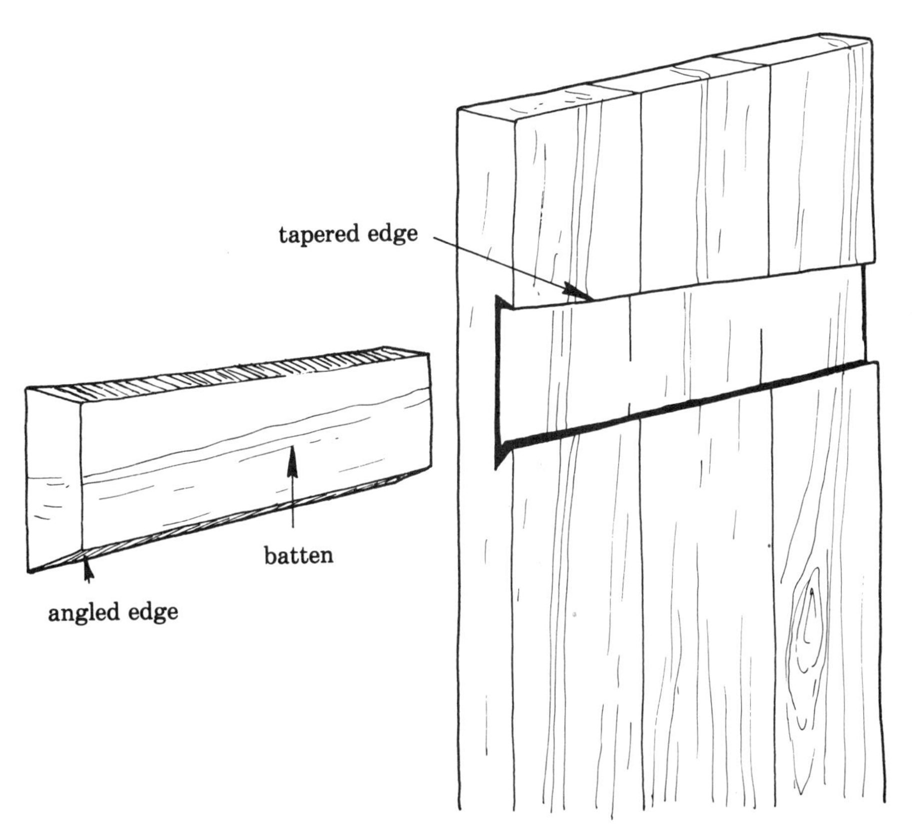

Illus. 12.4. ***Tapered sliding dovetail and batten detail (angles exaggerated).***

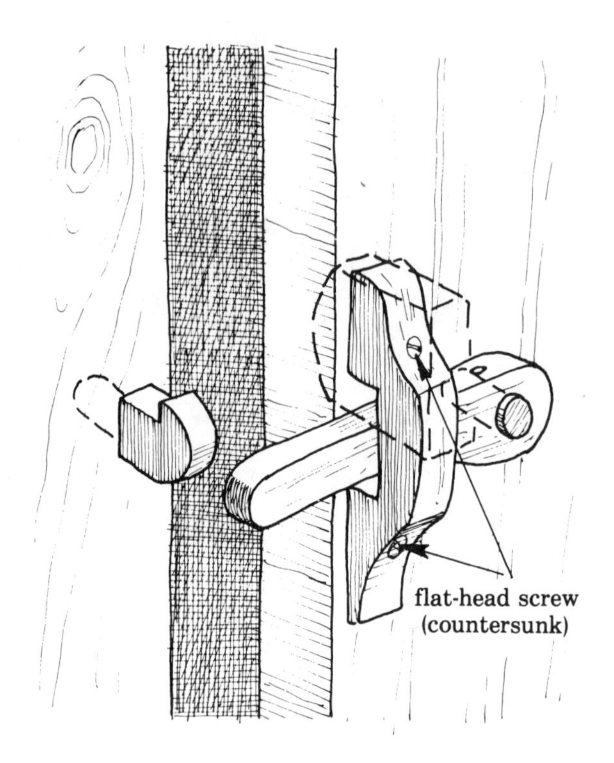

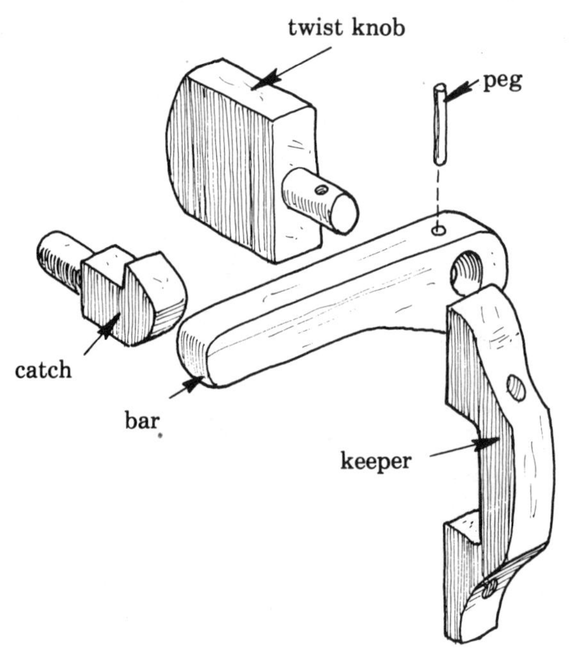

Illus. 12.5. *Door latch.*

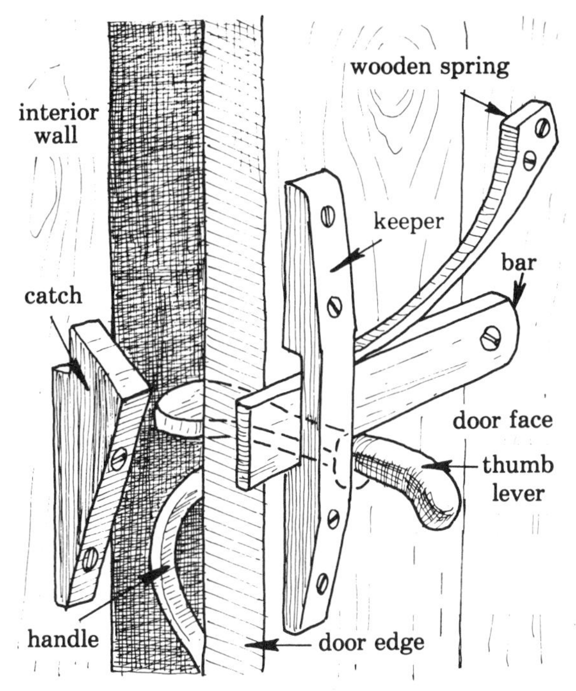

Illus. 12.6 *Spring latch with wooden spring.*

on the door, finish them to match the patch dowels.

The hardwood hinges for the doors can be cherry, hickory, white oak or beech. Like the door wood, they should be seasoned to a moisture content comparable to interior wood in winter. The batten section of each hinge should extend across all the boards of the door to provide additional support for the joints between the boards (see illus. 12.2). Saw the hinge pieces from 1¼-inch or thicker stock, and dress them with a hand plane. The batten usually tapers slightly from the hinge side to the latch side of the door, but the degree of taper isn't important. The best way to assure this is to drill the hole with a good drill press. Otherwise, use a framing square to align the drill points of your drill.

The pin on the pintel section of the hinge can be turned on a lathe, shaped with a hollow auger or carefully whittled. The pintel is the piece screwed to the door frame that supports the batten and door. The pin should be ¾ to 1 inch in diameter. Cut the pin parallel to the side of the pintel that will attach to the door frame to be sure it will line up with the hole in the batten.

Attach the hinge battens to the door by screwing through the boards into the battens. Carefully predrill and fasten at least three screws for each board. Countersink and fill all the screw holes.

Trim the sides, top and bottom of the door to fit inside the doorway. Saw and plane the edges to size. Allow 3/8-inch clearance along the bottom and 1/8 inch on the sides and top, but plane a little more off the latch edge at the side opposite the hinges to allow extra clearance as the door opens. Stand the door in place with small wedges driven between the door and the door frame. Make sure the pintels are plumb, one above the other, then use large wood screws to fasten the pintels to the door bucks. Countersink and plug the screw heads.

Another method of joining the boards is with sliding dovetail battens. These hardwood battens are inserted into tapered mortises cut into the inner or outer surface of the door after the boards have been butted edge to edge and glued. The battens slope about 3/8 inch per foot. The edges of the battens and mortises are also angled, about 5 degrees, so the battens can be wedged in place without glue.

Cut the angled edges of the mortises using the backsaw held against a wooden strip set at the proper angle. Make the perpendicular cuts, then

Photo 12.4 ***Wooden door latch made by Peter Gott.***

chisel out the waste wood. Cut the sliding dovetail battens slightly longer than the width of the door. Tap them snugly into the mortise, then saw off the waste at each end, flush with the door sides.

PROTECTING THE WOOD

If the growth of fungus can be prevented, weathering alone will only degrade unprotected wood at a rate of about ¼ inch per century. Log walls protected by generous overhangs will weather even less. Shingles and exposed trim, however, are more susceptible to weathering than logs and therefore require periodic maintenance or replacement.

Water entering corner notches, interfaces between logs and chinking, or between frames and logs can become trapped. During warm weather, these moist areas encourage the growth of decay fungus. To keep moisture out, caulk any gaps, and provide good ventilation in the crawl space and attic to allow accumulated moisture to escape. Notch joints and exterior window- and doorsills should, of course, slope downward to prevent rainwater from standing on the wood.

Insects, as discussed in chapter 4, Design, are another prime source of wood damage. Wood should never contact the earth directly, and even where it rests on the foundations you may want to treat it with preservatives, which, along with metal flashing, should keep insects from entering the wood.

WOOD PRESERVATIVES

In order to be effective, wood preservatives must penetrate the wood surface. The most effective applications are made under pressure, in a vacuum or with heat. Simpler methods include soaking, dipping, brushing and spraying. Soaking requires large amounts of material, special containers and considerable time. Spraying is not recommended, since all preservatives contain toxic substances (some are carcinogenic) that inevitably coat everything nearby.

Shingles and trim should be treated before assembly, but treating external surfaces has little value if moisture can get to untreated internal areas. Either treat the whole piece, or make sure all gaps are sealed.

Water repellents and water-repellent preservatives are used to stop or inhibit wood degradation due to fungi as well as insects, though good design and careful construction should minimize the need for such treatments. Many water-repellent preservatives contain a fungicide, such as pentachlorophenol. "Penta" is a highly toxic substance, which makes it too risky to be worthwhile. Two much safer preparations are Cuprinol 10 and Cuprinol 20. Cuprinol 10 contains copper naphthenate, which causes a green color that is supposedly not lasting. However, the color is still very apparent on our log ends two years after application. Clear Cuprinol 20 uses zinc naphthenate as a fungicide. Both products are advertised as nontoxic to humans, plants and animals *when dry*. Creosote is also effective, but the coal-tar base has been identified as a carcinogen. No water-repellent preservative should be used indoors.

Simple water-repellent treatments can be used to minimize the effect of weathering where a fungicide is not necessary, for example, in re-treating shingles or protecting trim and logs that are open to rain and weather. Water repellent made from paraffin wax, resin and solvent has been found by the U.S. Department of Agriculture Forest Products Laboratory to provide excellent protection against outdoor weathering for wood trim. Test samples were intact and free from decay after 20 years of exposure. The water-repellent treatment provided as much protection for trim as a commercial water-repellent preservative, suggesting that preservatives such as pentachlorophenol aren't required to protect trim from decay aboveground.

A basic water-repellent formula, for a 5-gallon batch, calls for a 1:3 ratio of boiled linseed oil and solvent (paint thinner or turpentine) and ¼ to 1 pound paraffin wax. Kerosene or diesel fuel can be substituted for the suggested solvent. The resulting mixture would be cheaper, though there would be some odor and a longer drying time. The less expensive mix could be used for roofing shingles or logs subject to rain. A higher-grade mix, for window sashes and doors, can be made by substituting tung oil for boiled linseed oil, with turpentine for the solvent and the minimum quantity of paraffin.

We immerse bundles of shingles into a 5-gallon paint bucket about two-thirds full. After ten seconds, we pull the bundles out and reverse them, dipping the untreated ends. Ten seconds later we set the bundles in a second bucket to drain. A piece of scrap wood holds the bundle off the bottom. The mixture flows between bundled shingles if it is not too cold. The only disadvantage of this treatment that we are aware of is drying time. Treated shingles remain oily until they have been on the roof for several days. Preservatives that we've made with the maximum paraffin content seem to be more effective as a water repellent, but they're even greasier.

To make these mixtures, melt the paraffin in a double boiler, and warm the solvent and boiled linseed oil to about 100°F. Stir the paraffin into the solvent. A batch not used within a few days may require warming and remixing. The treatment should be applied by dipping or brushing dry wood. Re-treat after one year. Further treatments every five to ten years are recommended.

Penetrating stains are water repellents or preservatives with the addition of one or more tinting colors in suspension. Penetrating stains are used to even or modify the color tone of exposed woodwork, in addition to protecting them from weathering, fungi and insects. The colorant also reduces degradation of wood by acting as an ultraviolet screen. One quart of oil-base colorant with 1/8 pound zinc stearate is used in a 5-gallon batch, along with the other ingredients for a water repellent or preservative.

Surface mildew removers can be used to lighten the color of stained logwork. We have used granular and liquid X-14, manufactured by White Laboratories, Inc., with excellent results. The Pratt & Lambert Paint Company suggests a homemade mixture consisting of 1 quart household bleach, 3 quarts warm water and 1 tablespoon powdered nonammoniated household bleach.

APPENDIX I

DRAWINGS

A few basic drawing supplies are a great help in getting practical, usable plans on paper. Start out with ¼-inch-grid graph paper. It's usually overpriced, but with graph paper and a pencil, you can start working up a set of plans to scale without the encumbrance of specialized drawing equipment. For the final drawings, you may want to use an architect's scale, a drawing board, a T-square and a 30°–60° plastic triangle. Quality pencils are also worth their price (HB is a good grade for general use). A $2 architect's scale bought at a shopping center is adequate for most work, but good quality paper is another story. Designer's layout paper, sometimes called bond, is available at stationery stores, as well as at engineering and art supply houses. I also recommend heavy tracing paper for copying the final drawings. Tracings can be blueprinted (check the Yellow Pages), and for a few dollars you can also order several sets of working plans. The advantage of having multiple copies of plans becomes obvious after a set gets rained on at the building site, or misplaced behind the seat of your pickup truck.

Several types of drawings are generally prepared: site plans, floor plans and elevation drawings. (Examples of the drawings I made for our house are shown in Appendix II.) Site plans show the area around a house, from above. Floor plans, also from above, show the layout of the building with the roof removed, one drawing per floor. Elevation drawings represent the house as seen from the exterior, with one drawing for each side of the house.

The site plan should be worked up before detailed construction drawings are tackled. This is a standard procedure for professional designers that is commonly glossed over by owner/builders. The area involved can be quite extensive, especially for woodland or homestead projects that may include roadways, adjoining forest areas, utility buildings and garden plots. A scale of ¼ inch to 4 feet allows sufficient detail, while permitting a view of a large area on a single sheet of paper. The site plan does not have to be an example of the surveyor's art, but a fair degree of accuracy, particularly in regard to proportions and relationships of various features, is very useful. Topographic maps can be used as a starter for locating major landmarks. These are available from map stores or the U.S. Government

Printing Office, and cover all areas of the United States. The remaining work can be roughly plotted with the help of a friend and a 50-foot tape measure. An inexpensive hand level and a sighting compass are very useful.

The site plan should include a general idea of contours, stream beds and wooded areas on the site, with north clearly marked on the plan. The house itself can be represented in simple block form. The site plan should indicate the fresh water source, any utility lines, waste systems, nearby buildings (standing or planned) and possibly a generalized schematic of ornamental plantings, orchard and garden.

General floor plans represent the layout of rooms, along with doors, windows, stairs and built-in fixtures or appliances. Special floor plans can be prepared showing joist and girder placements, furnishing and traffic patterns, plumbing and electrical installations.

Floor plans should begin as very rough sketches—drawings that are little more than doodles. Especially at the start, feel free to modify dimensions and move rooms about. Sketch in traffic patterns as your imagination wanders about the house. Traffic refers to people's movement in the house. For instance, you can design convenient places for people to set down packages and hang wet raincoats. Arrange rooms and stairways to minimize cross traffic and congestion in multiperson work areas. We included an exterior stairway to the upstairs. Storage areas should be conveniently located so that they are properly used.

Various arrangements can be experimented with on scale floor plans. To simplify making changes, make cardboard cutouts to scale of furnishings and appliances. For these games, you may want to use a large, 1-inch-to-1-foot scale. Otherwise, begin with ¼ inch to 1 foot. With this scale, a 16 by 20-foot pen will be 4 by 5 inches. More detailed drawings later should be done with a larger scale. For example, using ½ inch to 1 foot, ¼ inch equals 6 inches, a convenient scale for drawing wall thicknesses.

Draw doors with half circles to indicate swing direction and area. Indicate the swing of hinged windows also. In locating a stove or fireplace, be sure to consider requirements for safe installation. If you plan a 16 by 20-foot pen, decide if that means inside or outside dimensions—35 square feet of living space are at stake. Gradually add elements until you have a thorough understanding of how the many details relate to one another and how the house will be built and *used*.

Work up elevation drawings simultaneously. A staircase might seem conveniently located on the floor plan, but appear very much out of place on the elevation drawing. Also, keep in mind building details that are characteristic of log construction, such as log shrinkage and requirements for log notching and placement of joists and girders. Elevation drawings should include roof lines, doors, windows, flues and other external features. Exposed foundation work is also included. I like log-for-log elevations, since they give a feeling for the construction method. Drawing in the logs is helpful in understanding how the notched corners, joists and other details work together. It's usually impossible to know exact log sizes before actual construction, but a generalized drawing is very useful. I figure logs to be 1 foot in diameter, which is a very nice size to work with. For these drawings, I discount chinking altogether. Logs in the drawing appear as parallel horizontal strips, with the end view of adjacent corner logs represented as stacked rectangles (for hewn logs) or circles. Elevation drawings are especially helpful for working out details that could be confusing, such as foundation wall/sill log arrangements, joist notching, window placement and roof construction. A set of log-for-log drawings is also useful when it comes time to buck those long trees into house logs.

Elevation drawings should include allowances for log shrinkage. Green logs may shrink ½ inch or more per foot of log diameter. Ceiling height clearance beneath beams may drop as much as 6 inches during the first year or two while logs are drying out.

Take your time and mull over the construction process. When you have a thorough understanding of the full building, reexamine your thinking. Once construction begins, you may want to make further changes. This is understandable, as plans on paper only *represent* the full-scale, three-dimensional reality of a building. But successful changes during construction necessitate a thorough understanding of the complete structure. For instance, moving a door several feet could affect other details, such as the location of a joist or a woodstove.

Plans should also be coordinated with ideas for future additions. If you want to add a solar greenhouse, consider its location and shape in relation to the rest of the house. We designed a compact solar greenhouse located below our downstairs south-facing windows. Household light and our view were more important to us than a large-scale installation.

Roof type, size and orientation should be considered from an early design stage also. Gables generally span the narrow dimensions to simplify roof construction. Generous, well-planned overhangs will protect log walls and notched corners without cutting off natural sunlight, especially needed in winter.

APPENDIX II

DRAWINGS FOR LANGSNER HOUSE

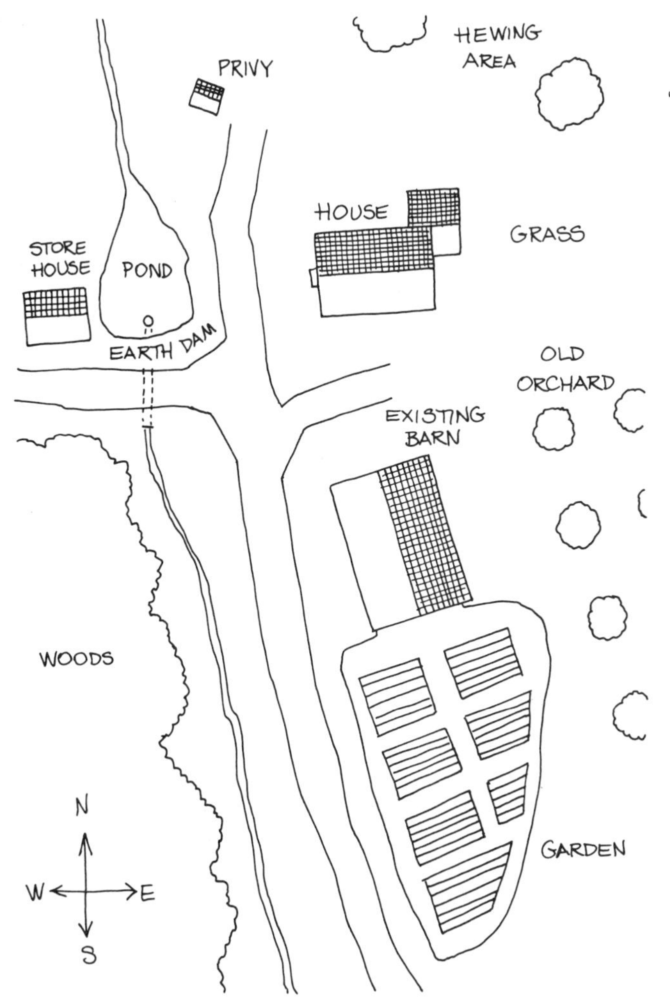

Illus. A.1. *A preliminary site plan of the Langsner house.*

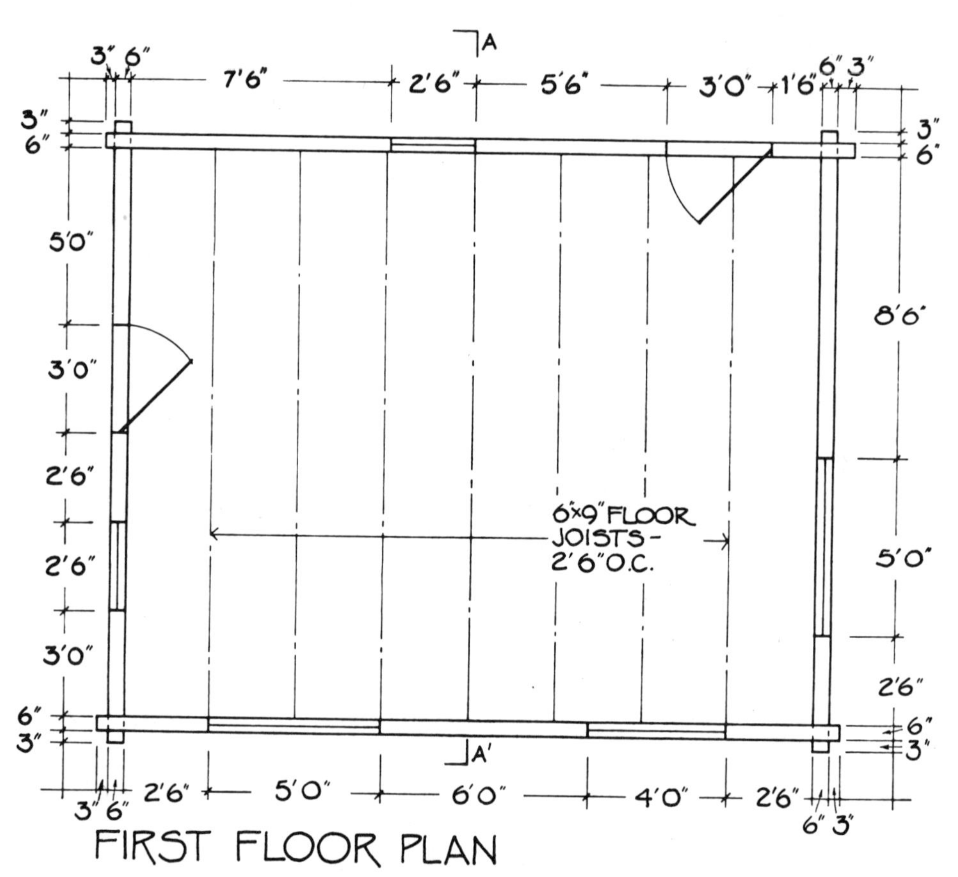

Illus. A.2. *Floor-plan drawing of first floor: Langsner house.*

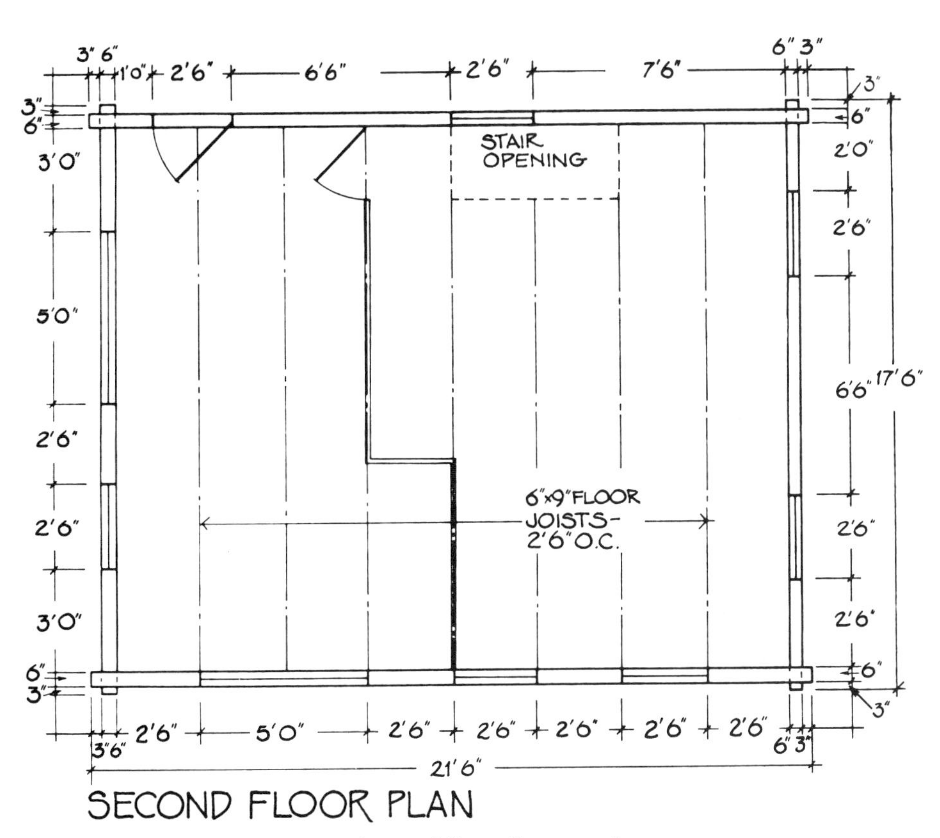

Illus. A.3. ***Floor-plan drawing of second floor: Langsner house.***

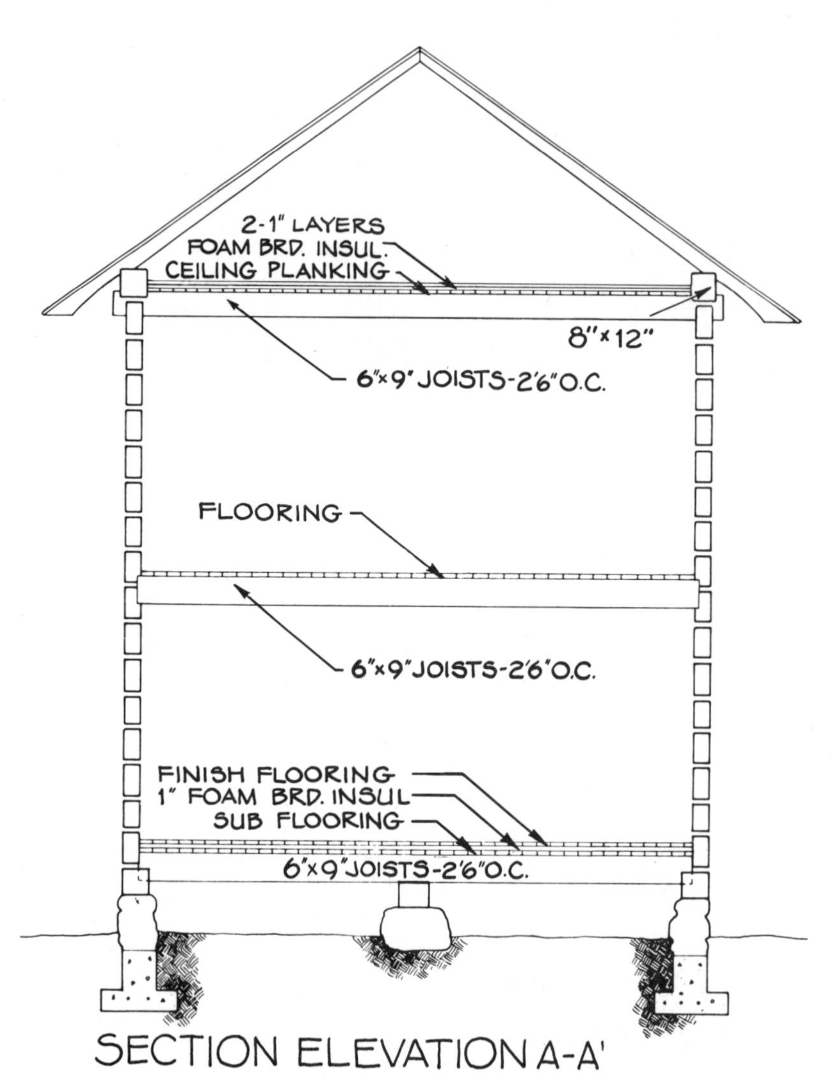

Illus. A.4. *Elevation drawing: section (see illus. A.2), Langsner house.*

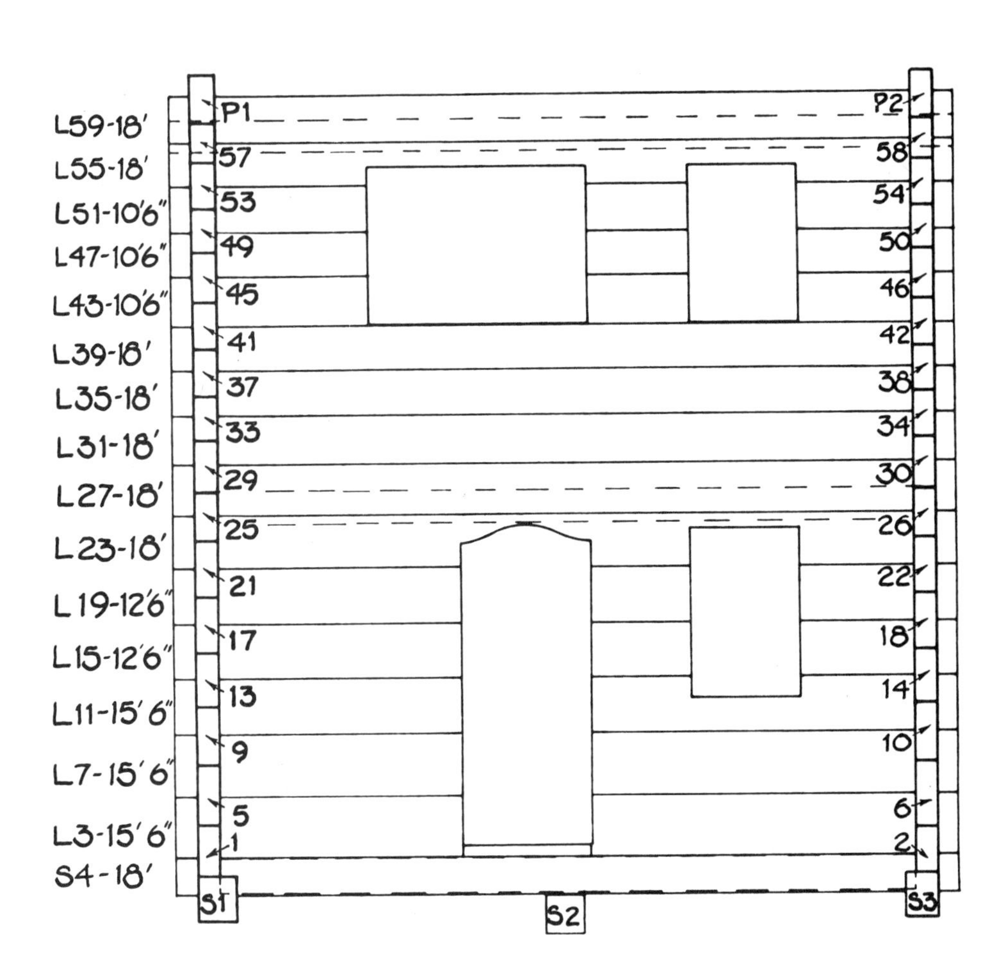

Illus. A.5. *Elevation drawing: west side of Langsner house.*

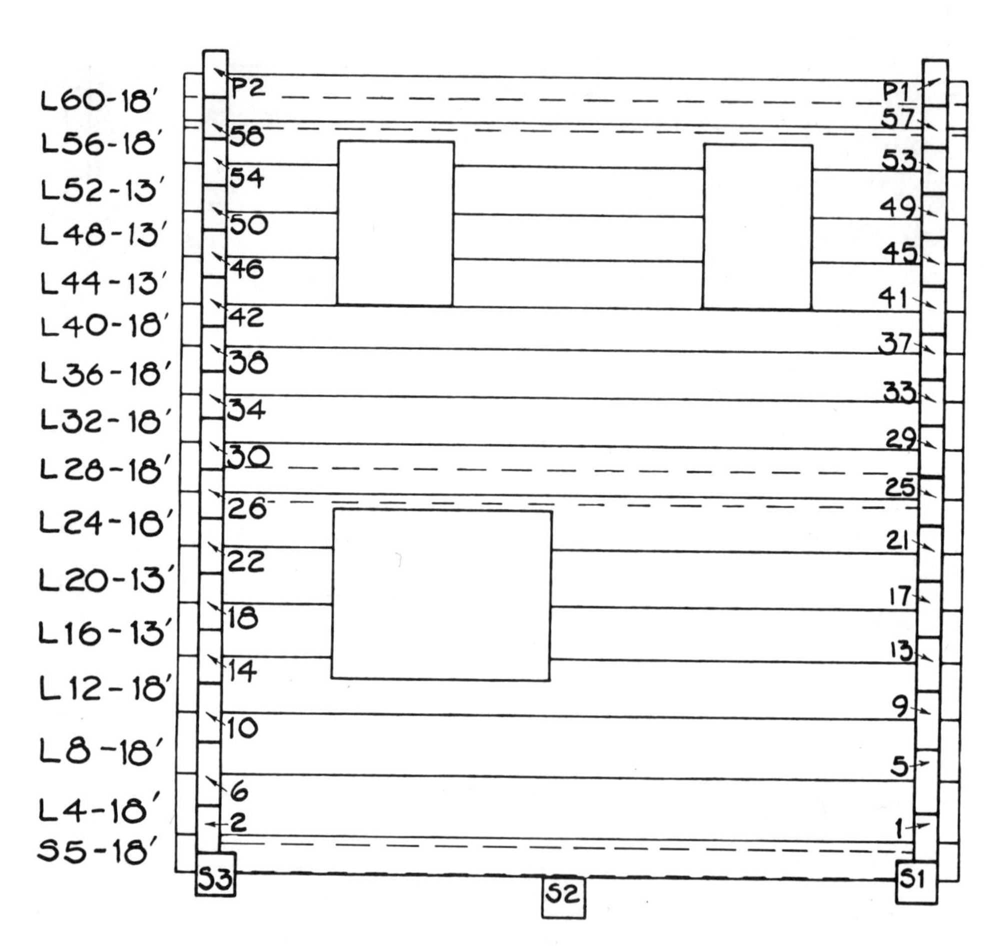

Illus. A.6. *Elevation drawing: east side of Langsner house.*

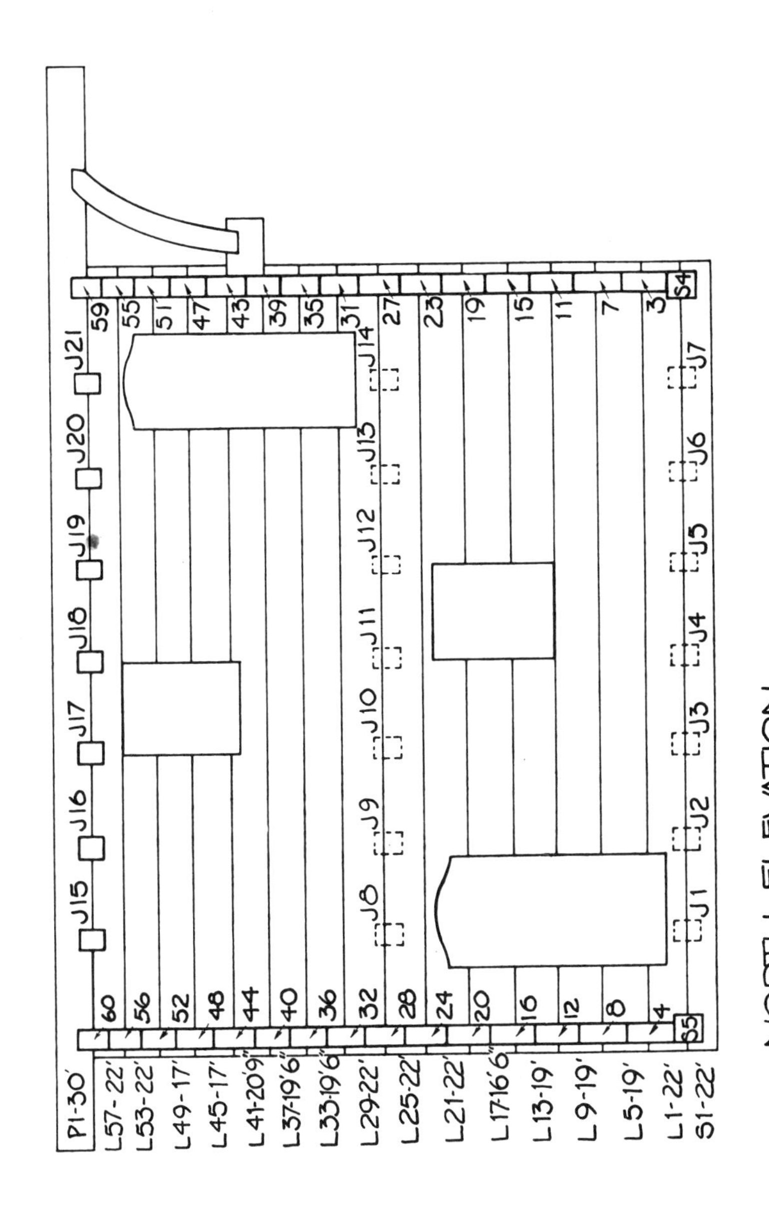

Illus. A.7. *Elevation drawing: north side of Langsner house.*

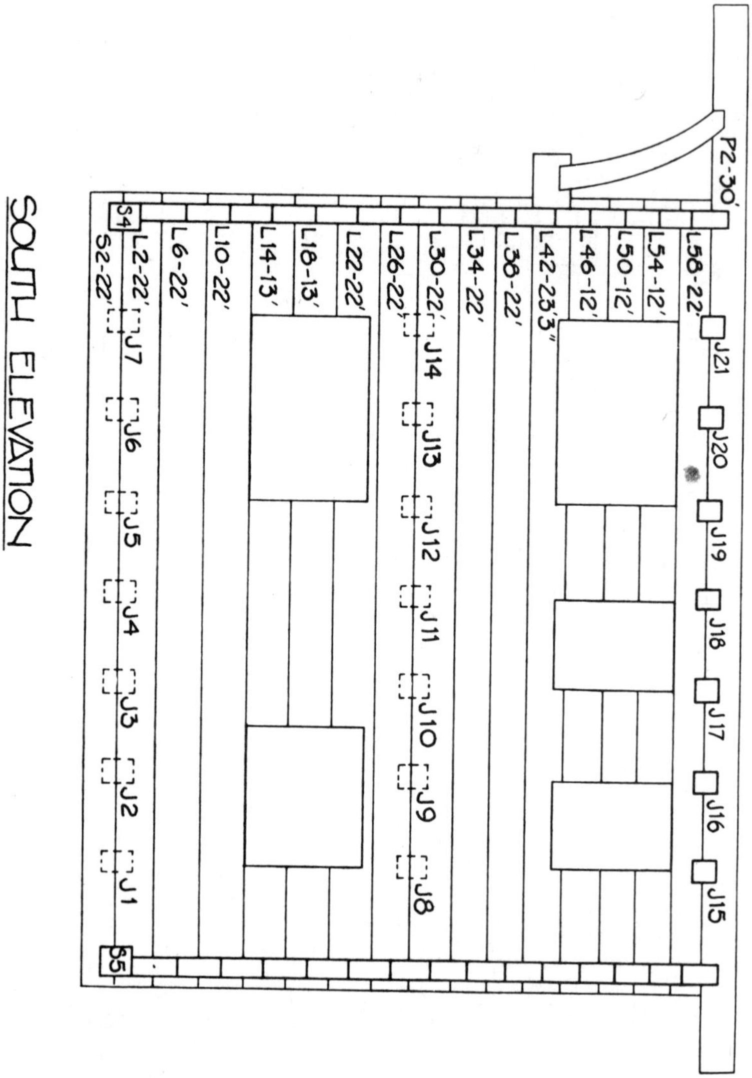

Illus. A.8. *Elevation drawing: south side of Langsner house.*

APPENDIX III

TOOLS A BASIC CHECKLIST

Site

- 50–100-ft. tape measure
- compass
- minimum/maximum thermometer
- shovel
- watch
- notebook

Design

- drawing board
- T-square
- architect's scale
- circle template (or compass)
- masking tape
- graph paper (¼-in. grid)
- bond layout paper
- tracing paper
- pencils
- eraser

General Carpentry

- tote box
- 5-pt. crosscut saw
- 16–20-oz. hammer
- pencil
- pocket knife
- 18–24-in. level
- line level
- notebook
- chalk line (blue)
- 20–25-ft. tape measure
- framing square
- combination square
- wrecking bar
- 1-in. chisel

Logging

- crosscut saw (or chain saw)
- saw-sharpening kit
- felling wedge
- axe
- plumb bob
- peavey (or cant hook)
- skidding tongs (or grab hooks)
- 100-ft., ⅜-in. cable with block
- tractor
- safety helmet, face and ear protectors
- timber crayon
- 20–50-ft. tape measure
- notebook
- pencil
- 12–16-ft., ⅜-in. logging chain (or ⅜-in. choker cable)
- rucksack

Foundation

- round-point shovel
- square-nosed shovel
- pick (or mattock)
- 5-ft. construction bar
- 50–100-ft. tape measure
- level (or transit)
- plumb bob
- builder's twine
- general carpentry kit
- axe
- wheelbarrow
- cement mixer (or mortar boat)

Hewing

- 2 (or more) trestles
- 2 notched crib logs
- crosscut saw (or chain saw)
- axe
- broadaxe
- 18–24-in. level
- 2 awls
- 2 chalk lines
- red and blue pencils
- tote box
- 20–25-ft. tape measure
- 12-in. stainless steel ruler, flexible
- hewing staples (or iron dogs)
- framing square
- straight-leg dividers
- pocket knife
- peavey (or cant hook)
- timber carrier
- timber cart

Notching

- hewing tools (except broadaxe)
- short, straight adz
- 2-in. chisel
- maul
- notch template (or 30°–60° triangle)

Raising

- peaveys (or cant hooks)
- timber carrier
- timber cart
- skid poles
- rope
- come-along (or block and tackle)
- 2-inch chisel
- mallet

Shinglemaking

- several iron wedges
- red and blue pencils
- straight-leg dividers
- splitting maul (or sledgehammer)
- axe
- shingle brake
- froe
- broad hatchet
- drawknife
- shaving horse
- froe club (maul)
- water-repellent preservative
- wooden glut

Chinking

- scaffolds
- ladder
- mortar boat
- mortar hoe
- margin trowel
- pointing trowel
- hawk (or mason's trowel)
- tin snips
- square-nose shovel
- buckets
- small scrub brush (to wet mortar)
- hammer
- tape measure
- gloves

APPENDIX IV

TOOL SUPPLIERS

Baileys, Inc., P.O. Box 550, Laytonville, CA 95454
(Chain saw and logging equipment; catalog $1.00)

Ben Meadows Company, 3589 Broad St., Atlanta, GA 30366
(Forestry and engineering supplies; catalog $3.00)

Blairhampton Alternative Resourcery, Inc., P.O. Box 748, Haliburton, ON K0M 1S0, Canada
(Logbuilding tools and classes; catalog $2.50)

Fine Tool Shops, Inc., 20 Backus Ave., Danbury, CT 06810
(Woodworking tools; catalog $5.00)

Frog Tool Co., Ltd., 700 W. Jackson Blvd., Chicago, IL 60606
(Woodworking tools; catalog $2.00)

Garrett Wade Company, 161 Ave. of the Americas, New York, NY 10013
(Woodworking tools; catalog $3.00)

Renovator's Supply, Millers Falls, MA 01349
(Reproduction hardware; catalog $2.00)

Silvo Hardware Company, 2205 Richmond St., Philadelphia, PA 19125
(Hand and power tools; catalog $1.00)

Tremont Nail Company, P.O. Box 111, Wareham, MA 02571
(Cut nails and reproduction hardware)

Woodcraft Supply Corp., 313 Montvale Ave., Woburn, MA 01888
(Woodworking tools; catalog $2.50)

Woodline, 1731 Clement Ave., Alameda, CA 94501
(Japanese woodworking tools; catalog $1.50)

Zip-Penn, Inc., Box 179, Erie, PA 16512
(Chain saw supplies)

APPENDIX V

FURTHER READING

LOGBUILDING

Mackie, B. Allan. *Building with Logs.* 7th rev. ed. Prince George, B.C., Canada: Log House Publishing Co., 1979.

——. *Notches of All Kinds.* Prince George, B.C., Canada: Log House Publishing Co.

McRaven, Charles. *Building the Hewn Log House.* Hollister, Md.: Mountain Publishing, 1978.

Suzuki, Makoto. *Wooden Houses.* New York: Harry N. Abrams, 1978.

FOLK BUILDING

Brunskill, R. W. *Illustrated Handbook of Vernacular Architecture.* New York: Universe, 1970.

Itoh, Teiji. *Traditional Domestic Architecture of Japan.* New York: Weatherhill, 1972.

Langsner, Drew and Louise. *Handmade.* New York: Harmony, 1974.

Rudofsky, Bernard. *The Prodigious Builders.* New York: Harcourt Brace Jovanovich, 1977.

MISCELLANEOUS BUILDING

Basic Construction Techniques for Houses and Small Buildings Simply Explained. Reprint. Prepared by the U.S. Navy. New York: Dover Publications.

Hylton, William H., ed. *Build Your Harvest Kitchen.* Emmaus, Pa.: Rodale Press, 1980.

Kern, Ken, and Magers, Steve. *Fireplaces.* Oakhurst, Calif.: Owner Builder Publications, 1978.

McCullagh, James C. *The Solar Greenhouse Book.* Emmaus, Pa.: Rodale Press, 1978.

Mazria, Edward. *The Passive Solar Energy Book.* Emmaus, Pa.: Rodale Press, 1979.

Shelton, Jay W. *Wood Heat Safety.* Charlotte, Vt.: Garden Way Publishing, 1979.

Stoner, Carol Hupping. *Goodbye to the Flush Toilet.* Emmaus, Pa.: Rodale Press, 1977.

Sussman, Art, and Frazier, Richard. *Handmade Hot Water Systems.* Point Arena, Calif.: Garcia River Press, 1978.

Wade, Alex. *Energy-Saving Houses.* Emmaus, Pa.: Rodale Press, 1980.

Wagner, Willis H. *Modern Carpentry.* South Holland, Ill.: The Goodheart Willcox Co., 1979.

LOGGING

Chain Saw Service Manual. Overland Park, Kans.: Intertec Publishing, 1976.

Dent, D. Douglas. *Professional Timber Falling.* Beaverton, Ore.: D. Douglas Dent, 1974.

Hall, Walter. *Barnacle Parp's Chain Saw Guide.* Emmaus, Pa.: Rodale Press, 1977.

Miller, Warren. *Crosscut Saw Manual.* Publication No. 001-001-00434-1. Washington, D.C.: U.S. Government Printing Office.

CRAFTS AND SKILLS

Adkins, Jan. *Moving Heavy Things.* New York: Houghton Mifflin, 1980.

Frid, Tage. *Tage Frid Teaches Woodworking,* Book 1. Newtown, Conn.: Taunton Press, 1979.

Kern, Ken; Magers, Steve; and Penfield, Lou. *Stone Masonry.* Oakhurst, Calif.: Owner Builder Publications, 1976.

Langsner, Drew. *Country Woodcraft.* Emmaus, Pa.: Rodale Press, 1978.

The Thatchers Craft. Wimbledon, England: Rural Industries Bureau, 1960.

Weygers, Alexander G. *The Making of Tools.* New York: Van Nostrand Reinhold Co., 1973.

WOOD

Black, John M. et al. *Forest Products Laboratory Natural Finish.* U.S.D.A. Research Note FPL-046. Madison, Wis.: Forest Products Laboratory, 1975.

Feist, William C. et al. *Protecting Millwork with Water Repellents.* Madison, Wis.: Forest Products Laboratory, 1977.

——. *Protecting Wooden Structures.* Madison, Wis.: Forest Products Laboratory, 1978.

Hoadley, R. Bruce. *Understanding Wood.* Newtown, Conn.: Taunton Press, 1980.

Rowell, R. M. et al. *Protecting Log Cabins from Decay.* General Technical Report FPL-11. Madison, Wis.: Forest Products Laboratory, 1977.

St. George, R. A. *Protecting Log Cabins, Rustic Work, and Unseasoned Wood from Injurious Insects in the Eastern United States.* Farmers Bulletin No. 2104. Washington, D.C.: U.S. Government Printing Office, 1973.

The Encyclopedia of Wood. Reprint of Forest Products Laboratory Handbook No. 72, *Wood Handbook.* New York: Sterling Publishing Co., 1980.

PERIODICALS

Fine Homebuilding. Newtown, Conn.: Taunton Press. Bimonthly.

Log House. Prince George, B.C., Canada: Log House Publishing Co. Annual.

New Shelter. Emmaus, Pa.: Rodale Press. Monthly.

INDEX